## Preface

When we see the uncountable stars in the night sky, we feel the vastness of the universe. We wonder how the universe come to exist, why the universe looks like the way we see. Was it like this forever or it came to this stage from something different, and what will be the destination of the universe? How and when the universe started its own journey? How the object would look like, if we could see those closely? Is there any life crawling, swimming and running out there? This book has been written with all those mysterious cosmic object that have been discovered so far. The pictures based illustration will reveal the mysterious universe for all ages.

*Note: Most of the pictures are taken from NASA website and other source with reference link. The more information can be found on the specific topics from those references.*

## About the author

Mohammad Sayed Khan was born and raised in a remote village of Bangladesh. Where the night sky seems like a flower garden of star in summer and autumn. He is passionate about cosmology and professionally he is an Aerospace Engineer. He was involved in developing space and satellite hardware in Ontario, Canada and currently working as a Research Engineer. This is his earnest effort to share his experience and understanding of the cosmos, its origin and evolution. Yes, Mohammad Sayed Khan, a tiny human creature hold the cosmos within himself, dedicated to know the origin and evolution of cosmos and our existence within it. We exist for a blink of eyes in a speck of dust called mother planet earth compare to cosmic period and vastness. Our ability to comprehend the cosmology is based on scientific method and effort made by all scientist and astronomer. In regard to this he has written few books about Cosmos for kids and adult and few more are in process.

# Mysterious Universe Revealed

## Contents

Introduction ..................................................................... 8
Is everything around the sun? ........................................ 9
Questions? ..................................................................... 10
What is the universe? ................................................... 13
How the universe started? ........................................... 16
    Planck Era ................................................................. 19
    GUT: Grand Unified Theory Era .............................. 20
    Electroweak Era ........................................................ 21
    Particle Era ............................................................... 22
    Era of Nucleosynthesis ............................................. 22
    Era of Nuclei ............................................................. 23
    Era of Atoms ............................................................. 23
    Era of Galaxies ......................................................... 24
    How did star and galaxy formed? ............................ 24
What was before the Big Bang? ................................... 28
How big the universe is? .............................................. 28
How big the solar system is? ....................................... 38
How big the sun is? ...................................................... 39
What is the biggest star? .............................................. 40
What is space and time? .............................................. 44
Where is the center of the universe? ........................... 48

What is our address in the vast universe? .......................49
What is the universe made out of?..................................51
What is the building block of matters?...........................53
What is dark Matter?........................................................56
How black hole is formed?...............................................57
What is Quasars? ..............................................................61
Why the universe is expanding? .....................................62
What is dark energy?........................................................63
How do we see our own galaxy? .....................................64
Can we live on another planet? ......................................66
How was the moon born? ................................................68
When did men go to the moon? ......................................69
Why does the U.S. flag on the moon have ripples?.........71
Can we live in mars? ........................................................72
Is there any other life in the universe? ..........................73
What is the time travel? ..................................................74
Can the child born in the space live in the earth?............78
What is gravity?................................................................79
What is black hole?..........................................................80
What happens when we go to black hole?.....................83
How Star is born and die?................................................83
What is Nebulae?..............................................................86
We are made out of star dust ? .......................................91
How our solar system born and evolve?.........................92
    What is solar system?..................................................95

- What we have in our solar systems?..........96
- Inner solar system..........98
- Outer solar system..........98
- Trans-Neptunian region..........99
- Asteroid Belt..........101
- Kuiper Belt..........102
- The Sun..........103
- How the sun makes energy?..........104
- What is the structure of sun?..........107
- What is Aurora?..........111
- What is solar flare?..........113
  - What is sunspots?..........114
  - What is solar wind?..........115
  - What is a solar prominence?..........117
- What is light?..........118
  - What is the interstellar medium?..........122
  - What is interplanetary medium?..........123
- What is in the Solar system?..........124
  - Mercury..........126
  - Venus..........126
  - Earth..........127
  - Mars - the red planet..........131
  - Asteroid..........133
- Meteoroid..........134
  - Jovian planets..........137

| | |
|---|---|
| Jupiter | 137 |
| What is escape velocity? | 139 |
| Why we are grateful to Jupiter? | 140 |
| Saturn - the ringed planet | 140 |
| Uranus | 142 |
| Neptune | 143 |
| Comets | 144 |
| Pluto and the other dwarf planets | 145 |
| Reference Data Summary | 145 |
| How sun effects our life? | 146 |
| What is Photosynthesis? | 148 |
| What is the cause of diversity? | 149 |
| Food chain or cycle | 152 |
| Ultraviolet Ray | 153 |
| The power of Sun | 154 |
| What is goldilocks zone? | 155 |
| Is there any life out there? | 157 |
| How do we know all this? | 159 |
| What is the International Space Station (ISS)? | 160 |
| How do astronauts live in space? | 161 |
| How old is the international space station? | 162 |
| How Big Is the Space Station? | 162 |
| Facts of International Space Station | 163 |
| What is the Hubble Telescope? | 164 |
| Hubble Facts | 165 |

What is the Large Hadron Collider (LHC)?......................167
How to map the universe? ................................................169
What is James Webb Telescope (JWST)? .........................171
Finally a Mysterious universe revealed!!! ........................172

# Introduction

The last few thousands of years human are wondering how big the universe is? Is there any edge of the universe? When we see uncountable number of stars in a clear night sky we wonder what is in the star, how do they look like, do they have any planet like us, and is there any life out there? With the development of different science faculty such as rockets, telescope, satellite and other physics, human now have the better visible access to the deep sky. Understanding the smallest particle of atom, now we know better about the mysterious cosmos. Even though the cosmos itself is vast but human does have much more knowledge about the cosmos ever before. We have our footprint on the moon 45 years back, we build international space station and people are living there, we are in the way of sending people in mars, we send satellite in every planet of our solar system, two of them named as horizon 1 and 2 are travelling outside of the solar system and are still active. Our understanding of cosmos are becoming rich with the knowledge of Astronomy, Cosmology, Astro-biology and Quantum mechanics .With the new research and new information collected from the cosmos, a new sets of question and challenges are developing. And scientist are interested and curious to reveal and resolve the new problem and challenge.

Before people believed that the earth is the center of the universe. The moon, sun, star in night sky all are revolving around earth. But with that wrong conception they found everything doesn't fit into this. Some object are going far, some are coming closer, some seems small, some big over the time. Our past conception has been proved wrong. We are very grateful to the scientist, engineer, and astronomer who are working to understand the universe with scientific prove and evidence.

## Is everything around the sun?

Around 3000 years from now, Babylon was the state of south Mesopotamia, which is current Iraq. Sumer and Akkad was two part of Babylon .The astronomy starts with some basic idea by Sumerians. The first written information was found about the position of planets, moon, and sun. 270 BCE Aristocrat, 4 BC Se- Shen, 140 AD Ptolemy, 400 AD Indian astronomy, 700 ad Arab astronomy evolved .In 1543 Copernicus revolutionize the idea that everything is around the sun and earth is one of the planet of sun and revolving around the sun .Then most of the people did not agree with him . As per them earth is the center of everything but he mathematically show them that sun is a star and earth is its planet .In 1610 Galilee Galileo see with his telescope that the moon of Jupiter is revolving around Jupiter. He proved Copernicus idea with observation.

We can't see the whole universe with naked eye. We see a small part of it. Even we can't see the distant object with telescope from the earth due to the earth atmosphere .So we setup Hubble telescope outside the earth atmosphere. And another very strong James Webb telescope will be lunch soon.

## Questions?

How far are the twinkle twinkle little star that we see in the night sky. How big are they? Hove you ever thought what is in the star? Were they looked like this forever? How they were formed, how long before? With naked eye we can't understand how big they are as they are far far away. You probably seeing a start that has been died before the light of that start reach to earth. Basically you see lots of star that are already died. All surprising events are happing in the universe. We in this science era we don't ignore them as an act of ghost. We want to reveal the mystery of those. So far we know so many thing that bring a lots answered which was never been answered before. But new findings also rises new question. Let's see how much you know about the universe where you live in. Do you know how the moon, earth, solar system are formed and how they are made of? How many small particles are found to break a material by which the material is composed of? What force is holding up those particles together? We all have the address, we have family. Where is our position in the vast cosmos? What is our cosmic address in the universe? And is our earth part of any family? What's outside of our earth? Is it just dark empty space? Is there anything in the darkness, some kind of strange thing that we cannot see? How big and how small we are? How big is our earth? How big is our sun than other star? What is the name of the largest star? How big is that? How big is the universe? How old the universe is? How it did born? What is the future of the universe? How our sun is formed in the universe? Were they like this all the time? Do the star and planet dies? Then how is it? How will our sun be destroyed? Can mankind ever be able to move to any other planet in the universe before it is destroyed? What is the time travel? Can the future be seen with time travel? Can I go in the past? Is time traveling possible? Are we alone in the universe? Or is there any intelligent life anywhere? Is there any life in the universe as it does have billions of billions of planets? Is there any other universe outside our universe?

In fact, thinking about how the universe formed and how it works is universal for human. In human civilization history we found lots of myth about the creation of universe. Almost all of them was based on imagination without any evidence and proof. But the modern science have been developed based on logic and reason and evidence and proof. In the last few decades new branches of knowledge have also been created such as astrophysicist, astro-biology, geology, paleontology, cosmology etc. Finally how much we know about the universe and how do we know? How do we prove so much? Let's see then.

The visible band of our Milky Way galaxy (Source: NASA)

Incredible mysterious universe is waiting for you to reveal. We are very small animal but out brain is intelligent enough that can hold the whole universe within us. To feel the vastness of cosmos you have to bring imagination with question. Science is the source of deep spirituality and to reveal the mysterious nature around us with reason, logic and evidence.

## What is the universe?

Cosmos is everything. What we can touch, what we can identify with our sensation even what we can't identify, everything is cosmos. Even you are a part of cosmos. Billions of star at night sky the full moon, earth, sun everything is part of cosmos. All are interconnected with invisible cosmic glue known as gravity. We don't know how big the universe but we can see a small portion without eye. It is like what you see from the shore of ocean. The shore of the cosmic ocean is event vast and far. Our universe is such a vast that you can't end of counting the star plant and moon. These are more that the number of sand grain that we have in our entire earth. Not only that cosmos is full of mysterious elements like nebulae, supernova, neutron star, black hole, galaxy and each one have some mysterious properties. There is big atmosphere surrounding earth and there is vast world with lots of mysterious items within it, some are very small and some are billion times bigger than our earth, and there huge dark cold empty space.

*"The Cosmos is all that is or was or ever will be. Our feeblest contemplations of the Cosmos stir us -- there is a tingling in the spine, a catch in the voice, a faint sensation, as if a distant memory, of falling from a height. We know we are approaching the greatest of mysteries."*

*— Carl Sagan, Cosmos*

The cosmos is made out of planet, star, galaxy, dust cloud, energy, light, even time and space. All are organized in a certain way. There was nothing but the energy before the Big Bang, not even time and space did exist. It is hard to imagine empty and infinite, we want to put boundary on everything. You might have question how the vast cosmos come into existence, how it formed, from where it started its journey, where it will end, even what it is made out of? In the cosmos there are billions of galaxy each of them does have billions of star. In between the galaxy is not empty, it is full of the cosmic radiation (light and heat), and magnetic field that can travel through empty space. The cosmos is unbelievable big. The fastest jet flight will take more than a million years to reach to the neighbor star of our sun. We know light travels 300,000 kilometer in every second. It will take 100,000 years to cross our Milky Way galaxy even at a speed of light. There are billions of galaxy like this in the known cosmos. Nobody knows the size of the cosmos as we can't see its edge but scientist can calculate the size of our visible universe .We can't see what is beyond the visible or observable universe as light from that part never reaches to us. The cosmos was not same as it is now, it is expanding very fast. Scientist develop a theory based on observation, evidence and proof that 13.8 billion years ago there was a mega expansion which formed space, matter particle and time starts it journey. As time pass space are formed by expansion as well as cosmos cool down and form matter from energy. Over the time it formed atom, stars, and galaxy. We will learn more about the cosmic structure like star and galaxy later.

You are related to everything around you. All the elements of your body came from the dead star. You are related to your favorite food, your favorite sports gear, green grass, spider, soil, water, and stone from moon, and rock from red mars, Saturn's spectacular ring, all are come from the same source. The water in your body is formed with hydrogen and oxygen atom. Do you know the age of the hydrogen? The answer is 13.8 billion years, yes 13.8 billion years back the hydrogen atom was form after the big bang. Rest of the atoms are cooked into a burning stars over the billions of years. When the star dies, its remnant left over was used to form our solar system, our earth, all life of earth including you.

**How the universe started?**

The universe as we see today was not like this before. 13.8 billion Or 13800 million years back the universe was different. At that time there was no life, no planet, no solar system, no star, no galaxy not even time and space. All the universe was squeezed into a point. This point like universe was very dense in energy and was very very hot. The whole universe was compressed into a point like pea seed, in a word all the energy was condensed into a point known as ground state energy. The universe starts its journey by rapid expansion from this point like seed. This is known as Big Bang. With the time space starts to form with rapid expansion. Time and space moving together, time passing and space is formed. Still the universe is expanding in all direction. The universe is not expanding into anything, nothing is holding the universe from outside. The universe is expanding into itself. There was no time and no space before big bang. Time starts is journey from the big bang. The point like seed universe was so hot, it was unable to hold such energy into a point. So it did expand rapidly and big bang happen. First few seconds was nothing but energy and this huge energy ball was unimaginable hot. Slowly it starts getting cool with expansion and some energy condensed into matter particle like electron, proton and neutron. Around 300,000 after those combined and form hydrogen atoms. Those hydrogen atoms and other ionized particle formed cloud. Yes, the cosmic cloud which is different than earth's cloud.

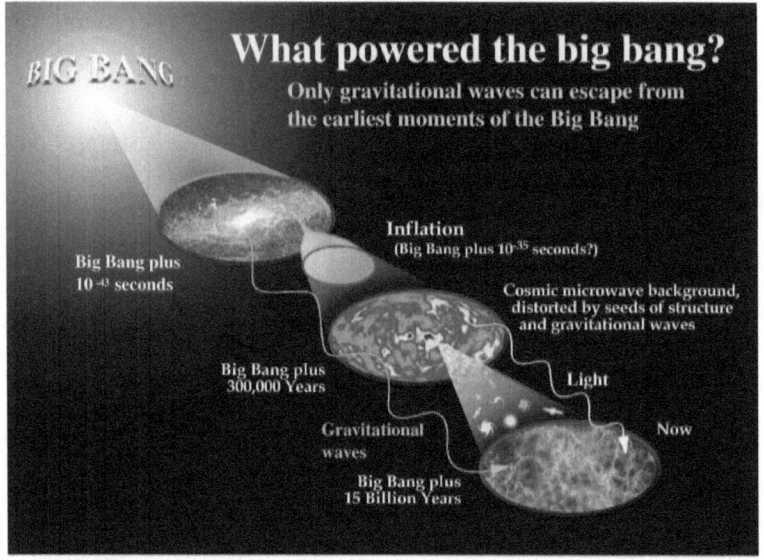

A short history of the universe (Image source: NASA)

The time line of the Big Bang is over 13.8 billion of years. Through advances in technology, we now have a better understanding how the Big Bang occurred. Big Bang Radiation Era after 300 thousand years of big bang has been detected. We can go back as far as planks era which is $10^{-43}$ seconds after the big bang. **Stars and Galaxies started to form after 1 billion years.** Our solar system formed 7 billion years after the big bang. From the big bang to present day scientist divide the time line in different era.

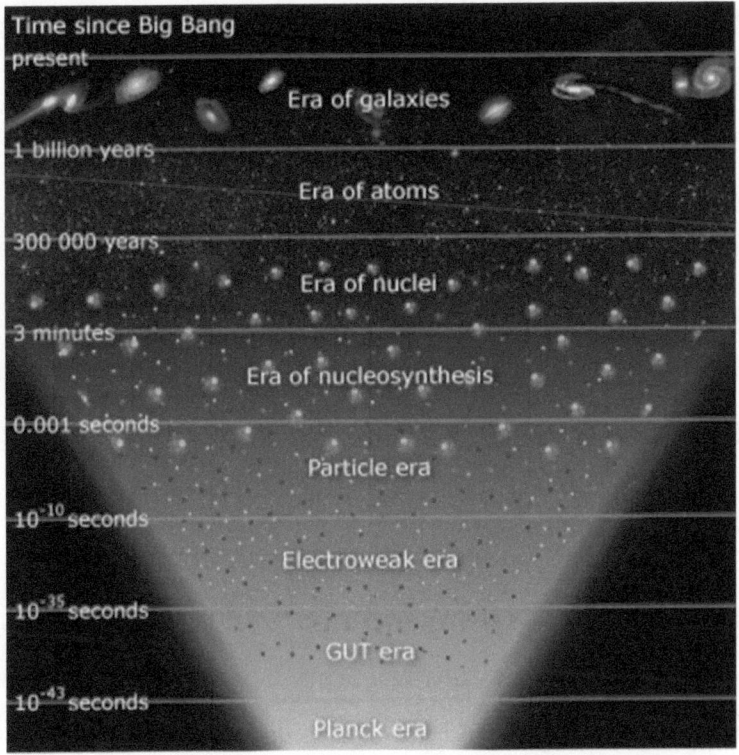

Differed Era or timeline after big Bang to present day
(Image: ESA)

**Planck Era** ($10^{-43}$ seconds)

The conditions were so extreme in the Planck Era that our current understanding of physics is inadequate to tell us much about it. Though physicists have a decent understanding of the early stages of the universe, the immediate fractions of a second following the Big Bang, known as the Planck Era, are not well understood. From the moment of initial expansion to $10^{-43}$ seconds afterwards, cosmologists suspect that the four fundamental forces that govern the universe today (strong, weak, electromagnetism, and gravity) were combined into a single unified force.

**GUT: Grand Unified Theory Era ($10^{-43}$ to $10^{-35}$ seconds)**
The Grand Unification Era followed by the Planck Era, taking place between $10^{-43}$ seconds to $10^{-35}$ seconds. The era began with the separation of gravity from the other three forces and ended with the separation of the strong force from the electroweak force. Under this conditions of extreme temperature and density, the four fundamental forces of physics looks more like different manifestations of a single force. Physicists are trying to developed a theory that unifies the strong, weak, and electromagnetic forces, called the Grand Unified Theory (or GUT for short), which includes the standard model of particle physics. Physicists would like to believe that gravity can be unified under the extreme conditions in the first $10^{-43}$ seconds, but so far this has not been demonstrated necessarily to be true from our current

understanding of the laws of physics. But the GUT Era ended when the strong force separated from the others, resulting in release of a huge amount of energy that caused the Universe to expand very quickly. As a result in a brief interval the Universe expanded by very rapidly. And from less than the size of a single electron to the size of a golf ball.

## Electroweak Era ($10^{-35}$ to $10^{-10}$ seconds)

At the beginning of the Electroweak Era, the strong force decoupled from the electroweak force, releasing a tremendous amount of energy and triggering a sudden expansion known as inflation. As space expanded more rapidly, extremely energetic interactions created elementary particles such as photons, gluons, and quarks. The era ended with the separation of electromagnetism from the weak force. The recent discovery of the Higgs boson suggests that our picture of the early Universe with the GUT era followed by the Electroweak Era is correct. The Higgs boson would be the only particle present during the GUT era. In the Electroweak Era, Higgs bosons could collide to create W and Z bosons that carry the electroweak force and quarks, which are fundamental to matter as we know it.

The Electroweak Era ended when the Universe cooled sufficiently that W and Z bosons were no longer being created; they decayed away and without them the electroweak force separated from the electromagnetic one and became the short-range weak nuclear force.

### Particle Era ($10^{-10}$ to $10^{-3}$ seconds)

At first, it was too hot for protons and neutrons to survive. Instead, there was a dense sea of "quarks" and "anti-quarks", the underlying particles out of which protons and neutrons (and their anti-particles) are made. There was a nearly equal mixture of quarks, anti-quarks and anti-particles in physics behave exactly as their counterpart particles but have opposite charge and annihilate if they collide with particles, so the mass of both the particle and anti-particle is converted to energy and emerges as gamma rays. As the Universe expanded and cooled, annihilation proceeded.

### Era of Nucleosynthesis ($10^{-3}$ seconds to 3 minutes)

The temperature remained high enough for the first 10 seconds that energy was still passed back and forth freely between electrons/antielectrons and photons. At this time, the conditions were at such high temperature and density that fusion reactions of protons into helium nuclei started. Hydrogen fusion, often called the proton-proton chain is shown to the left. The current-day abundances of hydrogen and helium support this.

**Era of Nuclei   (3 minutes to 300,000 year)**
Nuclear reactions could occur until about 3 minutes after the Big Bang. However, the temperature remained high enough to keep all the atoms ionized (electrons free from the nuclei) until about 300,000 - 500,000 years past the beginning. The change in absorption by the matter came about when the electrons combined with the nuclei, and only the photons at the specific line energies for the atoms were absorbed -- the Universe became mostly transparent, and the matter and radiation "decoupled". The matter was free to cool below the temperature of the photons, and the photon field no longer changed its properties through interactions with matter.

**Era of Atoms   (380,000 year to 1 Billion Year)**
The Era of Atoms began as the universe finally cooled and expanded enough for the nuclei to capture free electrons, forming fully-fledged, neutral atoms. Previously trapped photons were finally free to move

through space, and the universe became transparent for the first time. These photons have been passing through space ever since, forming the *cosmic microwave background*. The expansion since the origin of the universe has redshifted the initially energetic photons to microwave wavelengths. The CMB also marks the furthest point back in time we can observe – the time before is sometimes referred to as the *dark ages*. The differences in density seen in the CMB provided the seeds for galaxy formation. The first galaxies formed when the universe was roughly 1 billion years old and heralded the current *Era of Galaxies*.

**Era of Galaxies (1 Billion Years to Present day)**
The first galaxies formed when the universe was roughly 1 billion years old and we are at the current Era of Galaxies.

**How did star and galaxy formed?**
Edwin Hubble observe that the universe is expanding. Scientist estimate the age of the universe based on the rate of expansion. Galaxy formed just after a billion year from the big bang. Deeper into space you look, you see further back in time. So Hubble telescope can see the

young galaxies. There are two leading theories to explain how the first galaxies formed. The most popular one says that the young universe contained many small "lumps" of matter, which clumped together to form galaxies. Hubble Space Telescope has photographed many such lumps, which may be the precursors to modern galaxies. The galaxy-formation process has not stopped. Our universe continues to evolve. Small galaxies are frequently gobbled up by larger ones. The Milky Way may contain the remains of several smaller galaxies that it has swallowed during its long lifetime. The Milky Way is digesting at least two small galaxies even now, and may pull in others over the next few billion years. Our milky way will merge with our neighbor Andromeda in future as both are approaching each other. Galaxies are very massive, too, so their gravity is strong. When you crowd them together, the attraction can be so strong that two galaxies latch on to each other and don't let go. Eventually they merge, forming a single giant city of stars.

The cloud of hydrogen, helium and other ionized particle is known as nebulas. This huge, cold clouds of gas and dust is the birth places of star. Over a long period of time, the slightly denser regions of the approximate uniformly distributed matter gravitationally attracted nearby matter and thus grew even denser, forming gas clouds, stars, galaxies, and the other astronomical structures observable today. Wherever is more dense cloud is formed it start to shrink under their own gravity. The nebula begins to condense and form a ball. As the cloud gets smaller, it breaks into clumps. Inside of the clump eventually becomes very hot and dense due to external gas pressure. At some point the temperature and pressure reached so high that nuclear reactions begin. When the temperature reaches 10 million degrees Celsius, the clump becomes a new burning star.

Infrared observatories are able to detect heat coming from invisible stars that are forming inside such clouds. One of the most powerful of these is ESA's Herschel space observatory, launched in May 2009. Herschel will spend at least three years studying the dusty clouds where large and small stars are born.

**Simplified core collapse scenario:**

(a) A massive, evolved star has onion like layered shells of elements undergoing fusion. An inert iron core is formed from the fusion of Silicon in the inner-most shell.

(b) This iron core reaches Chandrasekhar-mass and starts to collapse, with the outer core (black arrows) moving at supersonic velocity (shocked) while the denser inner core (white arrows) travel sub-sonically;

(c) The inner core compresses into neutrons and the gravitational energy is converted into neutrinos.

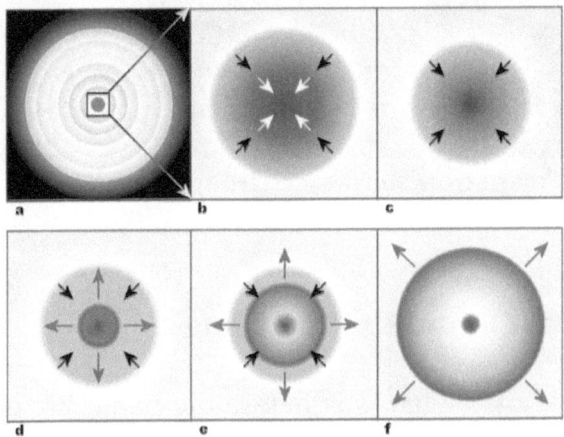

The star formation process Source
https://commons.wikimedia.org/wiki/File:Core_collapse_scenario.png

(d) The in falling material bounces off the nucleus and forms an outward-propagating shock wave (red).

(e) The shock begins to stall as nuclear processes drain energy away, but it is re-invigorated by interaction with neutrinos.

(f) The material outside the inner core is ejected, leaving behind only a degenerate remnant.

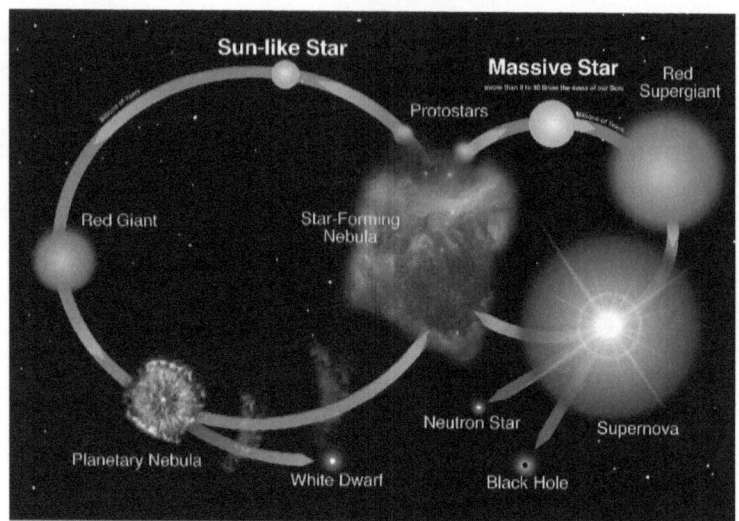

**Star formation sequence (Image Source: NASA)**

## What was before the Big Bang?

Nobody knows what was before big bang. We can go as back as back as planks time. Beyond that our laws of physics collapse. We know nothing beyond this time. Planks time is the smallest time that we can measure based on speed of light.

## How big the universe is?

Nobody knows how big the universe is. But the age of our universe is 13.8 billion years. As 13.8 billion years back everything was into a point and starts it journey by expansion. According to this, the diameter of the universe should be 27.6 billion light years. But we know the universe was expanding forever, the space is stretching and was not stretching at same at throughout the life. According to this our observable universe is much bigger. As per those information scientist found out observable universe is 93 billion light years in sphere. Remember we don't know the size and shape of our universe but we can see and observe a part of the universe that is 46.5 billion light years in radius or 93 billion light years in diameter. Beyond this sphere no light can comes to us. This is also known as light horizon as beyond this sphere the light web stretched to flat. As a result we can't see the light higher than the distance of 46.5 billion light years. A light years is the distance that light travels in a year. We know that light travels 1, 86,282 mile or 2, 99,792 kilometer in every seconds. So light can travel 5,878,625,373,184 mile or 9,460,730,472,581 kilometer in a year.

The speed of light is huge and nothing can go faster that the light. The time and space is being stretched or contracted to keep the speed of light constant. Our sun is 93 million miles away from us and it takes 8 minutes the sunlight to come to earth. So we can say our sun is 8 light minute away. Proxima centauri is the closest start to our sun and it takes 4.5 years to reach the light to us from there, so proxima centauri is 4.5 light years away from us. On the other way we can say that we are seeing 4.5 years old proxima centauri, if we want to see what is going on that start then we have to wait another 4.5 years to see the light of current moment. If something happen there or if proxima centauri dies we will not see it right away, we will see 4.5 years after. The further distance we look into space the oldest light we can see. If we can see deep enough we can see the light that emitted just after the big bang. Scientist can see the oldest light in a form of microwave and mapped this around our observable universe. This is known as cosmic microwave background radiation. (CMBR). A small number of temperature variation was observed which are considered as the seeds of galaxy structure.

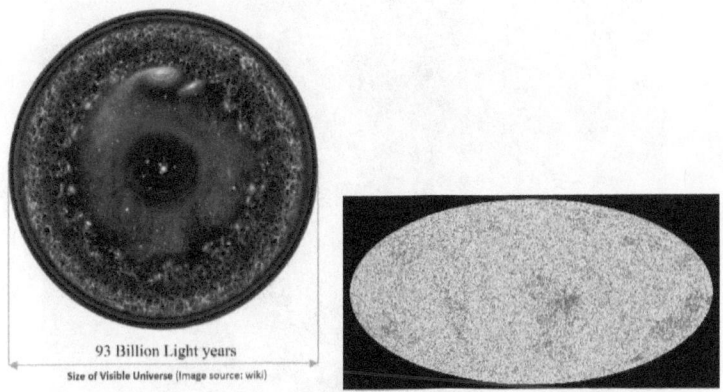

Cosmic micro web background radiation mapping of our visible universe (Image: NASA)

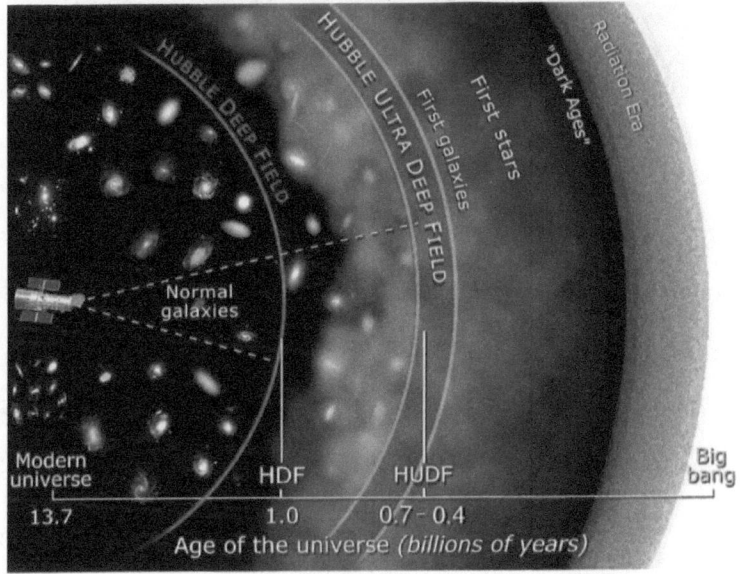

Our eye in space, the Hubble telescope deep field can see past by looking at deep field but it is not enough to see the baby universe, but James Webb telescope is coming soon to see that. (Image: NASA)

Cluster of galaxy by NASA Hubble Telescope

**Super Cluster:** Super cluster is the huge collection of galaxy. Our Milky Way galaxy belongs to Laniakea supercluster. This is a collection of 100,000 galaxies. It is stretched over 520 million light years. It consists of four sub parts as Virgo Supercluster (the part in which the Milky Way resides), Hydra-Centaurus Supercluster, Pave-Indus Supercluster and Southern Supercluster. Light years are used to express cosmic distance. In kilometer the distance will be so huge that it will be hard to image that number.

As radiation and particle travels almost at a speed of light, it is easier to measure everything is light years. Distance of one million parsecs is commonly denoted by the megaparsec (Mpc). Astronomers typically express the distance between the neighboring galaxies and clusters in mega parsecs. Distance of 3.262 light-years is commonly denoted by the parsec.

In cosmology galaxy filaments are considered as the largest cosmic structure. They are massive, thread like formation that is the boundaries between large voids in the universe and gravitationally bound galaxies. In a large scale most of the universe is empty space. Imagine galaxies are Light Island into a dark ocean. Spider like web form galactic filaments. It is thought that the dark matter dictates the galactic structure of the Universe. Galaxy that are made out of normal matter are cluster due to gravitational attraction. In the next smaller scale galaxy contains all the star system, nebulae, supernova and other mysterious object like black hole. In the next smaller scale star system like our solar system with its family exist. For an example out solar system with its family of planet, moon, asteroid orbit around Milky Way galaxy every 230 million years. This consecutive structure helps us understand the cosmic structure. This helps us to study it formation, size, etc. It is a way to see whole universe by top up or bottom down. If we break the normal matter then we will find molecule then further breakdown will reveal atom, event further will break to electron, proton and neutron. And even further will have lots of unstable subatomic particle and quark and lepton are the more stable fundamental particle that build all the matters. Quark and lepton can't be break further and scientists are working hoe energy condense into quark through the interaction. Scientist use microscope to see inside the matter and telescope to see the large scale cosmic structure. But we have limitation of microscope to see the quark as all as limitation of telescope. Moreover we live within the universe, so it is hard to see the whole universe at a time. But scientist can generate model with computer based on observation and theory. Our Milky Way galaxy would be looks like this.

The galaxy we live is called Milky Way Galaxy. Our neighbor milky way galaxy Andromeda is 780 kiloparsecs from Earth. Our solar system is within Milky Way galaxy and are situated in a local group galaxy cluster. Andromeda, Triangulum galaxy, few dozen dwarf galaxy into the local group cluster. The distance are shown in kilometer to feel the vastness of our neighboring galaxies .Our neighbor Andromeda galaxy are moving toward us and at some distance future both will be combined together to make a new spiral galaxy. All the neighboring object and their distance are shown in kilometers to realize the vast distance. But astronomer and scientist use light years to locate the distance of the cosmic object.

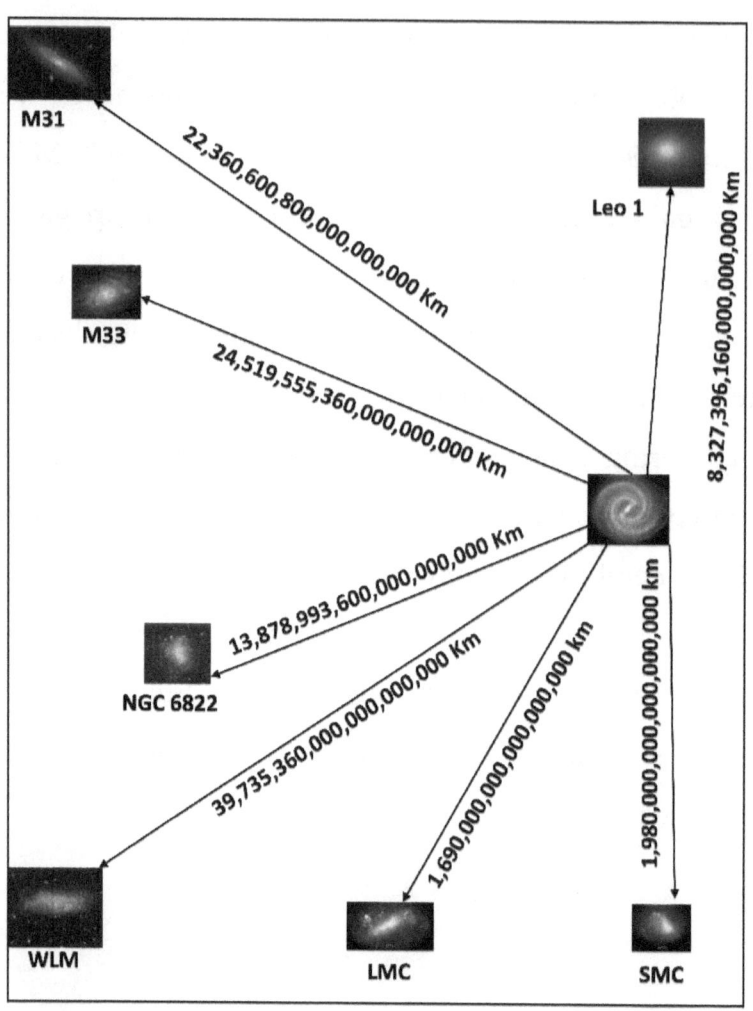

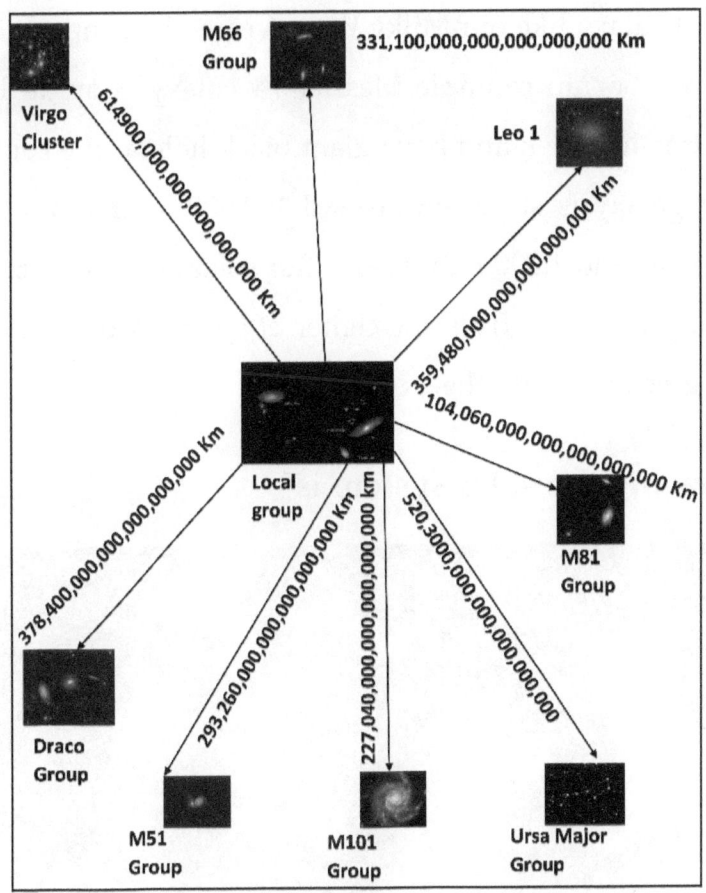

Neighbour galaxy from our Milky Way galaxy .Neighbour galaxy group from out local group cluster from our Milky Way. The distance are in kilometers. Image Source:NASA (https://imagine.gsfc.nasa.gov/features/cosmic/local_group.html)

It is impossible to take picture of our Milky Way galaxy as we live into it. But we know and see that our Milky Way galaxy is spiral type galaxy. By observing other spiral galaxy like our Milky Way, we get the shape of our galaxy. For an example Messier 74 galaxy is spiral like us. It is like disk and have giant black hole at the center. Our galaxy is stretched around 100,000 light years and thick around 1,000 light years. Our solar system is 26,000 light years away from the center of spiral. A disk can be imagine like a car wheel or DVD disk.

## How big the solar system is?

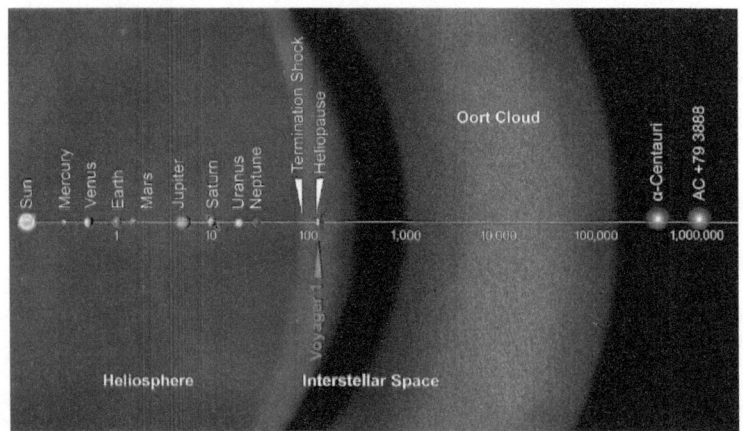

Our solar system is around 3 billion mile wide (Image: NASA)

Our solar system has eight planet, Neptune is the furthest planet. Pluto is a dwarf planet after Neptune. All are flooded with the plasma (Solar wind) that sun continuous bursting by burning its hydrogen fuel. Our solar system expand beyond the Pluto. The plasma particle expand beyond the orbit of Pluto. This flow of plasma particle known as solar wind. The boundary of solar system considered as far as the solar wind blow. This is known as halepause where the solar wind stopped. This is around 230 AU from the sun. The distance from earth to sun is 93 million mile and is known as astronomical unit or AU. This is used to express the Distance within solar system object. Voyager 1 and voyager 2 spacecraft cross the solar system boundary and still active in interstellar medium of Milky Way galaxy. Our closest neighboring star is proxima centauri that is 25 trillion mile or 4.5 light years.

## How big the sun is?

Sun is a huge ball of burning gas, this is very big compare to earth. The diameter of sun is 1,392,000 km or 109 earth will fit along the diameter. More than 99% of solar system mass are belongs to sun. If you can stand on sun surface then your weight would 28 times higher than the earth? Around 1.3 million earth will fit into the suns sphere. We see the sun small as we are far from it, it is 150,000,000 km from us. And light takes 8 minutes to come to earth from the sun.

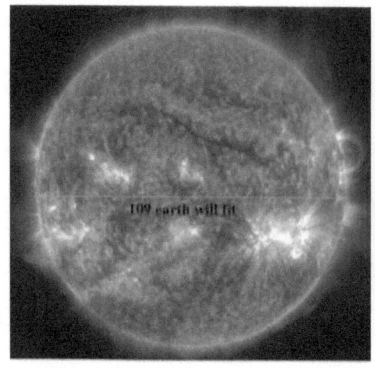 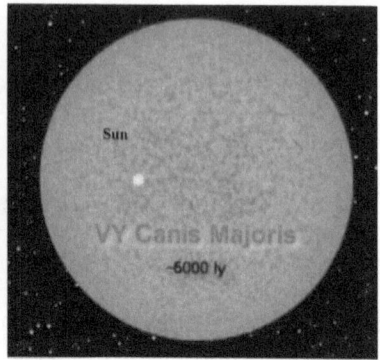

## What is the biggest star?

V-J Cains majoris is the known the largest star .This red supergiant is 4900 lightyears from the earth. 2000 sun can join together to make its diameter. It will take 1,100 years by an airplane to travel around this star whereas airplane would take only 32 hours to travel around earth's perimeter, 19 years around our sun.

It is hard to guess how big the moon by naked eye. Our moon's diameter is 27% of our earth diameter. Surface area of moon is 37.9 million square kilometer which is smaller than our Asia continent of 44.4 million square km. the whole earth surface is 5410 million square km so moon is around 7.5% of earth surface area. The volume of moon is 21.9 billion cubic km, seems a lot but our earth is around 1 trillion cubic km, so moon is 2% of earth volume. The mass of moon is $7.347 \times 10^{22}$ kg but earth mass is $5.97 \times 10^{24}$ means moon is only 1.2 % of earth mass. Around 50 moon can fit into earth.

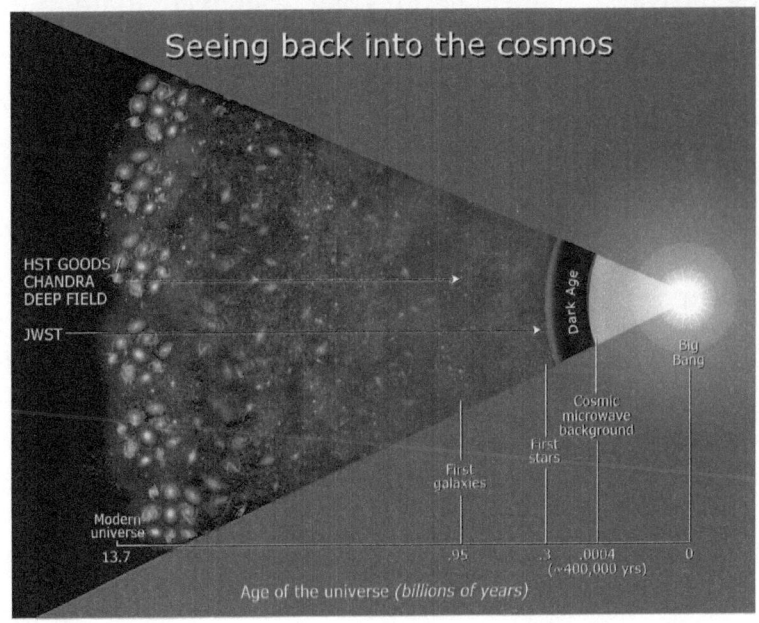

The average distance of the moon from Earth is 384,403 kilometers. Hubble telescope is located 570 kilometers away from the Earth. The new James Webb Telescope (JWST) will be set 1.5 million kilometer from the earth in 2019. The new James Webb will be able to see back in time.

The terrestrial planets in our solar system compare to the earth based on radius

Earth (6,371 km or 3,959 miles)

Venus (6,052 km or 3,761 miles) - 95% of the Earth's size

Mars (3,390 km or 2,460 miles) - 53% of the Earth's size

Mercury (2,440 km or 1,516 miles) - 38% of Earth's size

The last four planets of our solar system, size compare to the earth based on radius

Earth (6,371 km or 3,959 miles)

Jupiter (69,911 km or 43,441 miles) - 1,120% of the Earth's size

Saturn (58,232 km or 36,184 miles) - 945% of the Earth's size

Uranus (25,362 km or 15,759 miles) - 400% of the Earth's size

Neptune (24,622 km or 15,299 miles) - 388% of the Earth's size

Mercury is the smallest planet. Venus is similar to earth in size. Jupiter and Saturn are the giant in size. The Saturn ring is very wide but think. 17 earth can be fit along the Saturn ring. More can be learn from NASA website and Solarsystem.com.

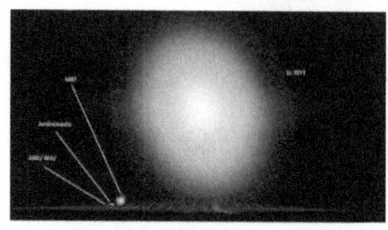

IC 1101 a supergiant elliptical galaxy of approximately 6 million light-years across, which makes it the largest known galaxy discovered to date. This is 320 megaparsec or 1.04 billion light years from earth. IC 1101 is more than 50 times the size of the Milky Way (which is some 100,000-120,000 light-years across) and 2,000 times as massive. If it were in place of the Milky Way galaxy, it would swallow up the Magellanic Clouds, the Andromeda Galaxy, and the Triangulum Galaxy.

## What is space and time?

Space spread everywhere. Imagine a place in a framework where everything has been set, such as planets, stars, football, and even you. The space can be expressed with three dimension, which is length, width and height. There is another dimension other than these three dimensions. It is time. At time dimension, we can go only one way, only at forward. Let's be clear some of our concept with the dimensions. In the language of physics, the dimension refers to any variable or any measurable value. There is no limit on the number of dimension. Of course it depends on what you want to measure. For example, color can be considered as dimension, temperature can be considered as dimension, weight can be considered as dimension. In space-time continuum there are four dimension. The spatial dimension that we express within three directions such as the length, the width, the height and the temporal dimension is the time as fourth dimension. Space-time constitutes a single framework in physics. This four dimensional concepts through which physicists have been able to solve many scientific complexities. With these four dimensional principles, the principles of particle can be explained that form the cosmic fabric of time-space where all the object is distributed.

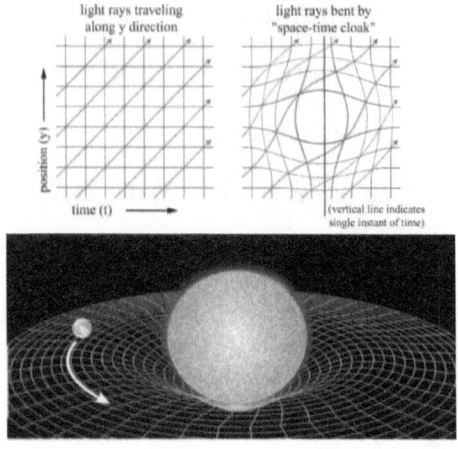

Einstein presented the space-time continuum with a new understanding of the cosmos- his theory of general relativity. In this theory, space is not an empty void, but an invisible structure called spacetime. Space-time is simply a two-dimensional grid through which matter and energy moves. Two dimensional grid of space-time is basically four dimension, three dimension of space and one dimension of time. It is hard to imagine with our brain. It is a four-dimensional structure whose shape is determined by the presence of matter and energy. Around any mass (or energy), spacetime is curved. The presence of planets, stars and galaxies deform the fabric of spacetime like a large ball deforms a bed sheet. This deformation occurs in four dimensions, so the two-dimensional bed sheet is a limited model. Try visualizing these depressions on all sides of a planet to build a more accurate image of this concept.

When a smaller mass passes near a larger mass, it curves toward the larger mass because spacetime itself is curved toward the larger mass. The smaller mass is not "attracted" to the larger mass by any force. The smaller mass simply follows the structure of curved spacetime near the larger mass. For example, the massive Sun curves spacetime around it, a curvature that reaches out to the edges of the solar system and beyond. The planets orbiting the Sun are not being pulled by the Sun; they are following the curved spacetime deformed by the Sun. This elegant way to express gravity was introduce by Einstein.

While you are reading this sentence is present. That happening this moment is present. The sentence you already read is past and next few sentence is open future. You are present is in between past and future. Off course this moment the word you are reading, more precisely the letter you are in is your present. What word or letter you have just finish reading is past. In other words, the time flows, your sense of feeling is constantly updating. Past experience are holding on your brain and the future gives the concept of the flowing period which is completely open. Your past can not be changed, and at the current moment you are experiencing is present and open future waiting.

According to Euclidean dimension Time is constant, no change in time. But as per the relative theory, there is no way to separate the time from the space. Because in this case, the time is based on the speed of the light, depending on the speed of the three-dimensional space. It also depends on the intense gravitational field, because the intense gravitational fields can slow down time. There is an integral mix of space and time in nature.

## Where is the center of the universe?

There is no center of the universe. From the beginning the universe was expanding in all direction. One galaxy is moving away from other galaxy. Imagine you are inflating a spotted balloon, as the balloon expand the spot on the surface of the balloon will move away from each other. Similarly the galaxy are moving apart. Universe is not expanding into anything, it is expanding into itself. As it expand space is formed, there is no space outside the universe. We can see the oldest light as farthest we can see.

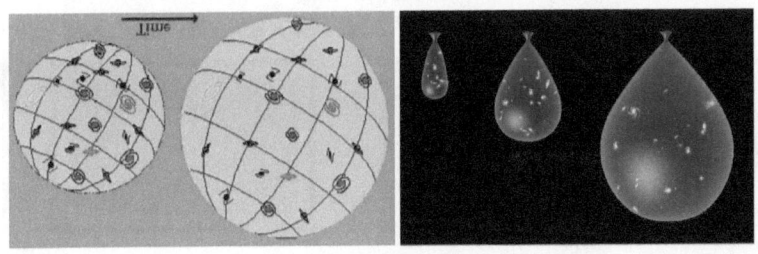

## What is our address in the vast universe?

There are more than 300 billion star in our Milky Way galaxy. One of the star is our sun. Our Milky Way galaxy is a spiral shaped galaxy. You can image the shape of pin wheel or when snake sleep by making a coil shaped. Our sun is 2600 light years from the center of the galaxy. Our galaxy is in local group cluster which belongs to Laniakea supercluster. More than 40 galaxy together formed local group cluster that is 10 million light years wide. Local group is part of Virgo supercluster and Virgo is part of Laniakea supercluster.

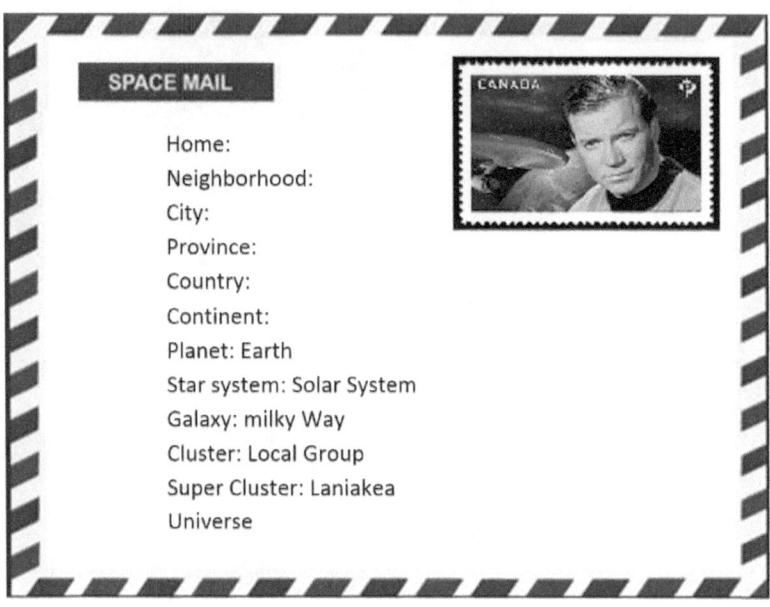

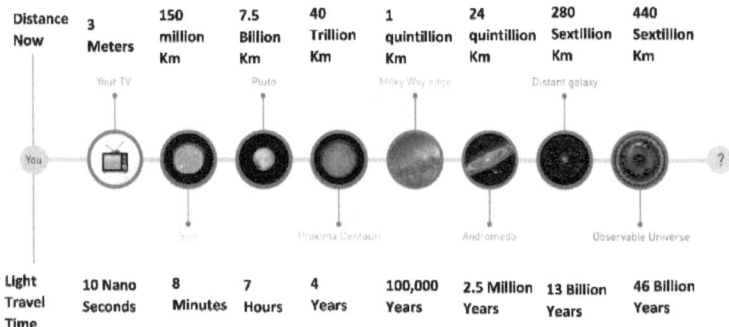

Image Source: NASA

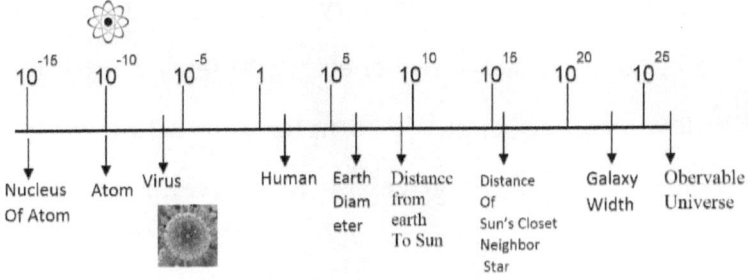

Our Relative Size on Meter scale

## What is the universe made out of?

This is one of the most basic questions humanity is asking forever. But it is one of the most difficult to answer. In ancient world people thought the five elements such as water, air, fire, earth and ether (sky) are the basic constituents. But our knowledge of matter and energy has progressed rapidly over the last few centuries. This help us to understand nearly all of the materials and phenomena around us. Scientists have even begun to understand the principles that have shaped the evolution of our universe itself over the past few billion years. This was Einstein and other scientist who open our eyes to understand the cosmos through a new look. The time, space, energy and matters are the four fundamental ingredient of the cosmos. Yes the time and space form time-space continuum by which Einstein explain the theory of relativity. The mass equivalence formula (E=mc²) show the energy condensed into mass. The flow of energy and it complex interaction with all these four ingredient governs the universe through a certain laws of nature.

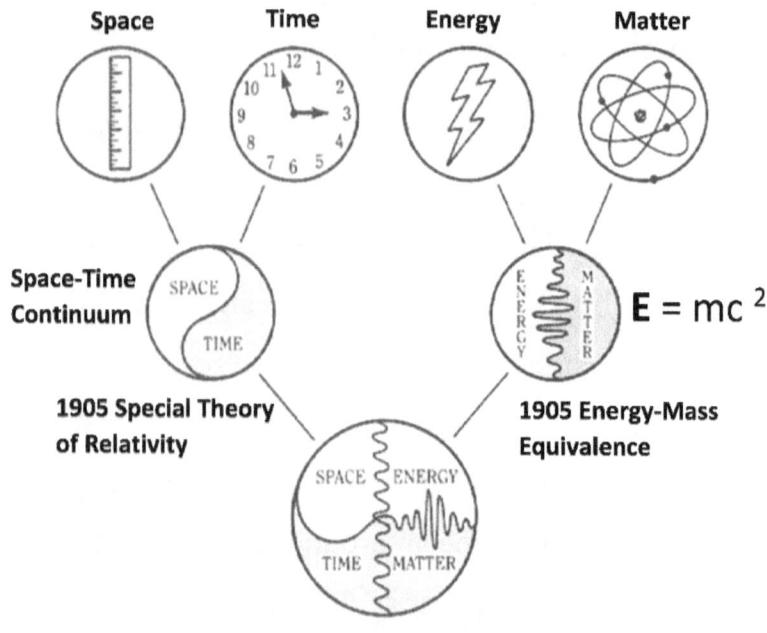

1915 General Theory of Relativity

The Universe is thought to be consist of three types of substance: 'normal matter', 'dark matter' and 'dark energy'. Normal matter consists of the atoms that makes up the stars, planets, human beings and every other visible object in the Universe. Normal matter almost certainly accounts for the smallest proportion of the Universe, somewhere around 4%, dark matter is 21 % and dark energy is 75%

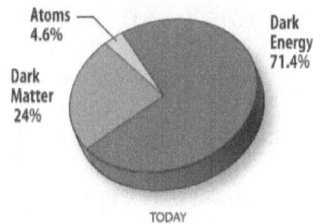

## What is the building block of matters?

Everything around you is made out of mater, all the object made out of material or matter. All matters are made out of different atoms and different atom combined to form molecules. The object or matter has mass and occupies the place. Mass is the total amount of an object that is constant all time. There is a difference between mass and weight. The weight is the force by which the earth is pulling, such as your weight. If we take one kilogram of sugar to the moon, then its weight will be different, but there will be no change in the total amount of the mass. While the gravity is very low in the moon, one sixth of the Earth's gravity. So weighing one kilogram or 1000 grams of sugar will weigh 165 grams at moon. Matter interact with light or the electromagnetic radiation. Matter are spread everywhere in the universe, all planets, stars, galaxies are made out of matter.

**Normal Matter:** Breaking the object or the matter can be found molecules, two or more atom combine together to make molecules. By breaking atom we will find Electrons, protons and neutrons. All atoms are formed with Electrons, protons and neutrons, but different number of Electrons, protons and neutrons makes different types of atom. Such as one electron and one proton makes one hydrogen atom. Two electron, two protons and two neutrons make helium atom. As those number increase atoms get heavier such as uranium. Proton and neutron composed of 6 types of quarks and Electron is one type of leptons out of six types. The quark and lepton cannot be broken anymore. There are four types of fundamental force act to build all matters in the universe. All Visible matter is found in atomic form such as planets, stars, stones, and people. In addition to this visible object or Matter there is another type of matter that we cannot see because they do not spread any light. That is call unknown matter or dark matter.

The four fundamental forces are gravity, electromagnetic, strong nuclear force and weak nuclear force. These four forces govern the whole universe. 12 fundamental stable particle (six types of quark and six types of lepton). These fundamental particle makes all matter by interacting with the four forces.

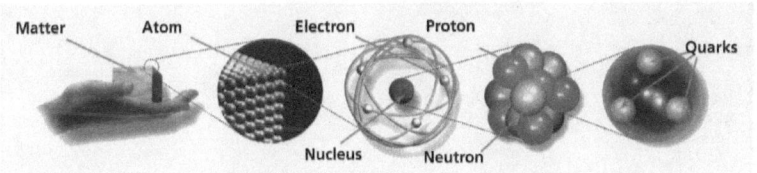

Image Source: CERN

The normal matter is in the form of an atom. 20 percent of the total matter in the universe is normal or baryonic matter. And the remaining 80 percent is the Dark Matter which we cannot see. The empty space of sky is not full of dark matter. First, it is dark, meaning that it is not in the form of stars and planets that we see. We are able to detect baryonic clouds by their absorption of radiation passing through them. Observations show that there is far too little visible matter in the universe. But high concentrations of matter bend light passing near them from objects further away, but we do not see enough normal matter that could bend that much. This suggest that such objects to make up the required 25% dark matter contribution.

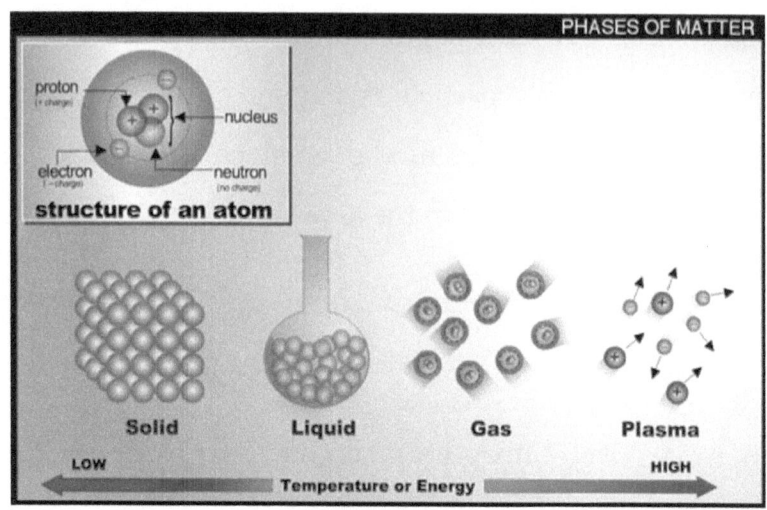

Structure of atom and different phase of matter with temperature (Image: NASA)

## What is dark Matter?

Scientists have figured out the amount of mass required to keep those things together, the amount of energy needed for the medium. It is to be said that the gravity is an invisible glue, so that you are on the surface of the earth, if there would be no gravity then you would not be able to stay with earth, you would be through from earth the surface into space. If there is no gravity, the earth would not rotate around the sun, there would be no structures, and everything all would have been shattered. Scientists have added all the matter and found just a little bit of mass and a significant mass is missing.

Most of these missing mass are Dark Matter. Since it does not emit light, it cannot be seen. Visible Normal Matter is very little in the universe. Dark matter is not antimatter, because we do not see the unique gamma rays that are produced when antimatter annihilates with matter

## How black hole is formed?

Primordial black holes are thought to have formed in the early universe, soon after the big bang. Stellar black holes form when the center of a very massive star collapses in upon itself. This collapse also causes a supernova, or an exploding star, that blasts part of the star into space. Scientists think supermassive black holes formed at the same time as the galaxy they are in. The size of the supermassive black hole is related to the size and mass of the galaxy it is in. A common type of black hole is produced by certain dying stars. A star with a mass greater than about 20 times the mass of our Sun may produce a black hole at the end of its life.

Stars are held in perfect balance by two opposing forces. There's the inward pressure of gravity, attempting to collapse the star, counteracted by the outward pressure of the emitted radiation. At the core, millions of tons of hydrogen are being converted into helium every second, releasing gamma radiation. Nuclear reactions in the core of the star produce enough energy and pressure to push outward. For most of a star's life, gravity and pressure balance each other exactly, and so the star is stable. As the star consumes the last of its hydrogen, it switches to the stockpiles of helium that it has built up. After it runs out of helium, it switches to carbon, and then oxygen. Since the star continues to pump out radiation, it balances out the gravitational forces trying to compress it. Stars with the mass of our sun pretty much stop there. Not massive enough to continue the fusion reaction, beyond oxygen, they become a white dwarf and cool down. But for stars with about 5 times the mass of our sun, the fusion process continues further up the periodic table to silicon, aluminum, potassium, and so on, all the way to iron. When a very massive star exhausts its nuclear fuel it explodes as a supernova. The outer parts of the star are expelled violently into space, while the core completely collapses under its own weight. If the core remaining after the supernova turn into neutron star if it is massive then turn into black hole.

And so, in a fraction of a second, the radiation from the star turns off. Without that outward pressure from the radiation, gravity wins out and the star implodes. An entire star's mass collapses down into a smaller and smaller volume of space. When a star runs out of nuclear fuel, gravity gets the upper hand and the material in the core is compressed even further. The more massive the core of the star, the greater the force of gravity that compresses the material, collapsing it under its own weight.

The velocity you would need to escape from the star goes up, until not even light is going fast enough to escape. From the perspective of the collapsing star, the core compacts into a mathematical point with virtually zero volume, where it is said to have infinite density. This is called a singularity. Where this happens, it would require a velocity greater than the speed of light to escape the object's gravity. Since no object can reach a speed faster than light, no matter or radiation can escape. Anything, including light that passes within the boundary of the black hole -- called the "event horizon" is trapped forever.

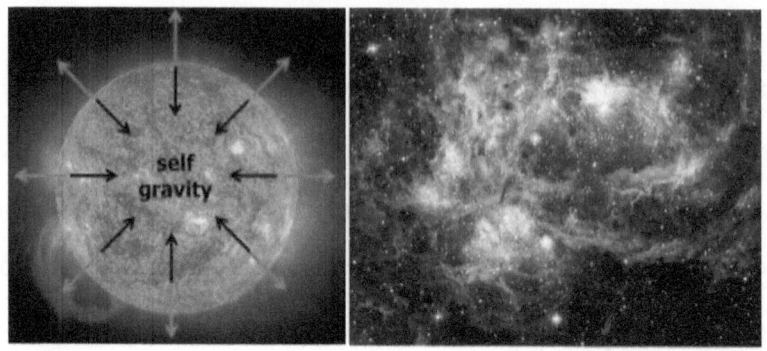

For small stars, when the nuclear fuel is exhausted and there are no more nuclear reactions to fight gravity, the repulsive forces among electrons within the star eventually create enough pressure to halt further gravitational collapse. The star then cools and dies peacefully. This type of star is called a "white dwarf." Our sun and one day will turn into white dwarf. There is nothing to fear that the amount of sunlight that is in the sun will last for another five billion years.

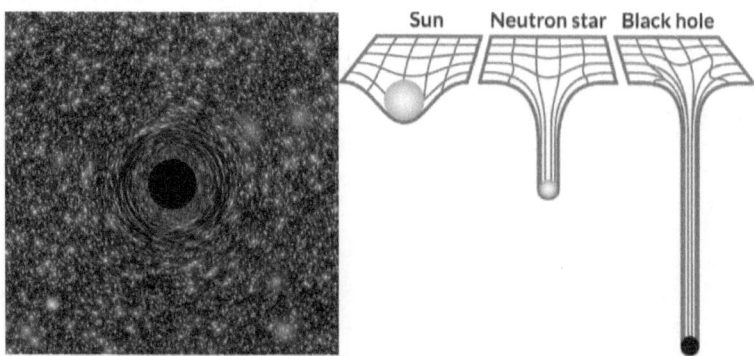

Black hole in spacetime curvature

This computer-simulated image shows a supermassive black hole at the core of a galaxy. The black region in the center represents the black hole's event horizon, where no light can escape the massive object's gravitational grip. The black hole's powerful gravity distorts space around it like a funhouse mirror. Light from background stars is stretched and smeared as the stars skim by the black hole. (*Source: NASA, ESA*)

It shows how different objects affect the space-time curvature. The strange object of the universe is black hole which is basically another condition of a large dyeing star. Whose one tea spoon of matter is a few trillion tons weight?

Artistic view of quasar of left and right quasar from earth

### What is Quasars?

Quasars is Very distant bright objects in our known universe. Its light is so bright that it looks very bright despite the light coming from very distant. The quasars are actually discovered by radio signals. After the discovery of radio telescope, scientists get a strong radio signal from a distant source. Then the Hubble Telescope was focused toward that direction and found this cosmic stranger. The very active and young galaxy has huge black hole at the center and a strong jet of light passes perpendicularly. This light is released from the event Horizon of the black hole. We now know that all galaxies have supermassive black holes at their centers; some billions of times the mass of the Sun. When material gets too close, it forms and accretion disk around the black hole. It heats up to millions of degrees, blasting out an enormous amount of radiation. The magnetic environment around the black hole forms twin jets of material which flow out into space for millions of light-years. This is an AGN, an active galactic nucleus.

## Why the universe is expanding?

Isaac Newton invented the gravity. He shows how the gravity holds everything together. Here's an interesting mystery is hidden. If gravity pulling everything in then why is not everything going to be crushed and collapse into nothing? Our universe is a dark ocean and galaxies are floating like an island into it. Gravity is holding our earth structure, Earth and other planets are revolving around the sun, the structure of our solar system, the Milky Way galaxy's structure are all for the invisible cosmic glue known as gravity. What is the force against this gravity which keeps everything in balance? The mystery is dark energy. Albert Einstein face the same problem as Newton. He mistakenly used the cosmological constants in his equation. He did not know that the universe was expanding. In 1929, Edwin Hubble saw in his telescope that the distant galaxy was moving away from us quickly. He proved that the universe is expanding. Galaxy is the collection of billions of stars that are rotating with respect to the center of the galaxy. Our sun revolves around our Milky Way galaxy with an average speed of 828,000 km / hour in orbit with his family. It takes about 230 million years to revolve in an orbit. It is known as galactic year.

## What is dark energy?

The universe is expanding from the beginning of the birth. Scientists are still researching how the universe is growing. The power of the mysterious energy, which is spreading faster than the invisible glue gravity is called the dark energy. This energy works against the gravity in a large scale cosmic structure. As the current model of the universe is thought to be 70% invisible unknown energy or dark energy, 25% dark matter and 5% normal matter. No one knows what dark energy is, but no doubt about its existence. We clearly observe its effect. Scientists are researching this, maybe its secret will be exposed in future.

## How do we see our own galaxy?

Imagine you see your city or town through your house windows. Same way we see our own galaxy. In the dark night sky we see our Milky Way galaxy. Because we are inside the Milky Way, we don't get chance to take any pictures of it from an angle "above" the Galaxy. We can see clear whole picture of our neighboring beautiful M31 (the Andromeda Galaxy). Imagine trying to create a map of your house while confined to only the living room. You might peek through the doors into other rooms or look for light spilling in through the windows. But, in the end, the walls and lack of visibility would largely prevent you from seeing the big picture. Similarly scientists collect the many different wavelengths of light and other information to build the map of our own galaxy. Our presence in the cosmos is very short. Like a blink of flashlight in the dark ocean. The Milky Way galaxy you see in a dark night sky, I am also watching the same one from cold night of Canada. May be someone from a seashore in Australia, someone from desert of Arab or Sahara, someone from remote village of Bangladesh, some little cute from Africa are watching the same sky. It is the same stars and galaxy all human are watching generation after generation. It is the same Milky Way, same sky our ancestor did watch with wonder and their curious imagination made dream and lots of folklore around the world. But we are lucky to have science like tools that reveal the mysterious sky to us including its story.

The main tools used by astronomers are telescopes, spectrographs, spacecraft, cameras, and computers. Astronomers use many different types of telescopes to observe objects in the Universe. Some are located right here on earth and some are sent into space. Scientist send satellite outside the earth surface to see deep into the sky or to see the oldest light of the universe.

## Can we live on another planet?

45 years ago people walked into the moon. Since then, space exploration is using by robots. The international space station is build and people are living there. NASA is planning to send people to Mars. It will take about two years to go to Mars. Astronaut must live on the red planet for a year to find the closest distance for return. Astronauts will need to bring all the food, water and other resources needed for the 6- to 9-month journey to Mars. Their bodies will need to be able to withstand the reduced gravity and increased radiation in deep space. Mars gravity is one-third of the Earth's gravity. There our weight will be one-third of the weight that we feel on earth. We will feel very light there. You can go higher by jumping over mars surface. The baby born in Mars will grow taller. There is no technology to live on another

planet yet, but scientists are doing lot of research on it. We build international space station above the earth and there are people living in the International space station and many more experiments are being conducted from there. NASA's human Mars mission presents even more challenges of sending humans safely to a farther distance and to a more dangerous environment. Designing an aircraft that can safely enter and exit Mars' unpredictable atmosphere is a big challenge. The two Mars Exploration Rovers launched and has successfully flown and land on a very precise spot of mars land. Both flown through about 483 million kilometers (300 million miles) to reach mars.

## What happened during the launch of the probe on Venus?

Venus atmosphere is very dense. The air pressure on the surface of Venus is equal to the pressure below the Earth's ocean under 800 meters. Carbon dioxide in Venus creates greenhouse effect. This is why we are very concern about the carbon pollution of earth. This bad carbon di oxide kills the protective ozone layer of earth atmosphere. Venus is the hottest planet in our solar system. When the satellite probe descended on Venus, its equipment melt and electronics were in short circuits.

When descends further down to the clouds its acid completely destroyed the probe. But scientists did not stop sending satellites on Venus, even after a dozens of satellites were successfully sent to Venus. In the future, scientist are dreaming to make a huge balloon space station above Venus.

## How was the moon born?

About 4530 million years ago, a huge object fell on Earth at the speed of 25,000 kilometers per hour. As part of this clash, some parts of it were left on the earth and the rest spread to space. Some parts of this material were pulled by the gravitational force of earth and begin to rotate around the earth. The moon is very important for the earth's life. The moon has a great influence on the climate of the Earth, such as tidal. As a result of this collision, the vertical axis of the earth moved a little bit. As a result, we are getting season change which is necessary for the emergence of diverse life.

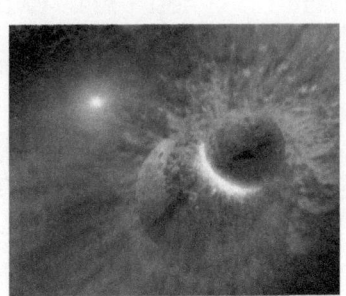 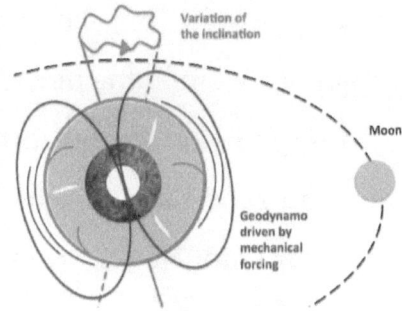

Moon Formation 4.53 billion ago and its tidal effect on earth

## When did men go to the moon?

Apollo 11 blasted off on July 16, 1969. Neil Armstrong, Edwin "Buzz" Aldrin and Michael Collins were the astronauts on Apollo 11. Four days later, Armstrong and Aldrin landed on the moon. They landed on the moon in the Lunar Module known as Eagle. Collins stayed in orbit around the moon. He did experiments and took pictures. On July 20, 1969, Neil Armstrong became the first human to step on the moon. He and Aldrin walked around for three hours. They did experiments. They picked up dirt and rocks of moon and left their footprint on the moon the two astronauts returned to orbit, joining Collins. On July 24, 1969, all three astronauts came back to Earth safely. Humans had walked on the moon, Walking through the moon was a matter of great pride for the entire human race. In history Collins is the loneliest people in the universe. When his two companions were in the moon's land, he waited for them to return to the moon orbit. Michael Collins, the Command Module pilot who was orbiting the Moon in the mother ship. He was waiting to take his fellow crew members home to Earth. During that time he was the loneliest person across the quarter-million miles of empty space from the Earth. Have you wondering what that would feel like?

Men on the Moon 50 years ago

**Why does the U.S. flag on the moon have ripples?**

Since there is no wind or atmosphere on the Moon, how can the US flag be flapping in pictures of the first Moon landing? Here on Earth, flags are pushed out by the wind and wind makes the flag flap. Obviously, there's no wind on the Moon, so what's holding the flag up? The answer is pretty easy. There's a rod, sort of the like a curtain rod running across the top. So the flag on the Moon is being held out by the rod and isn't blowing in the wind. This L shaped rod also made folds in the flag. The astronaut had to wiggle the flag pole to get it set into the ground properly. That wiggle induced a wave in the flag that continued for quite a while. Both effect together seems it looks like it is flapping.

## Can we live in mars?

Mars's atmosphere is extremely thin, mostly carbon dioxide which is poisonous for humans. Mars is colder than earth. But there are ways to turn Mars into a warm and breathable atmosphere. To convert it livable we first need water. If under the ground does not have enough water, then comet that are mainly compose of water ice can be divert to hit Mars. This will create ocean of water and cloud. And this thick atmosphere of cloud will hold the sun's heat and it will become warm. Then the algae from the earth can be send into the ocean .so this will create oxygen and release it to the atmosphere. Slowly the air will be breathable. This is the same process how our earth become breathable. This is called terraformation. But in reality, there is a lot of money and labor required to do this. May be one time people could do it successfully. Today's dream is tomorrow's reality.

## Is there any other life in the universe?

The chemical composition of Earth-based life is primarily derived from a selected few ingredients. The elements hydrogen, oxygen, and carbon account for over 95% of the atoms in the human body and in all known life. Of the three, the chemical structure of the carbon atom allows it to bond readily and strongly with itself and with many other elements in many different ways, which is how we came to be carbon-based life, and which is why the study of molecules that contain carbon is generally known as "organic" chemistry. The study of life elsewhere in the universe is known as exobiology, which is one of the few disciplines that, at the moment, attempts to function in the complete absence of first-hand data.

Not all the star have rocky planets. One in nearly 200 stars in the world has rocky or rock-like planets, where there is a possibility of life. Who knows there might have intelligent creatures like us or even better. If one of every one billion stars has civilization of intelligent creatures, then our Milky Way galaxy itself will have more than 300 places have possible civilization.

$$N = N_s \times F_p \times F_l \times F_i \times L_c/L_s$$

**N** is the number of civilizations in the Milky Way today.

**$N_s$** is the number of stars in the Milky Way.

**$F_p$** is the fraction of stars with habitable planets.

**$F_l$** is the fraction of habitable planets with life.

**$F_i$** is the fraction of life-bearing planets where intelligent civilizations arise.

**$L_c$** is the typical life-time of a civilization in years.

**$L_s$** is the typical life-time of a star (10 billion years for Sun-like stars).

In 1961, Frank Drake tried to make an estimate of possible civilization with his controversial single Drake equation.

## What is the time travel?

To define any place we use three dimensions which are length, width, and height. Further easier way to say, the place can be expressed in three dimensions: north-south, east-west and top-bottom. Anywhere we go, our position can be expressed by this dimension. The universe begins with the Big Bang, the space is being expanding since then. The space cannot be separated from time. Time is considered as the fourth dimension or the time-space goes together. Gravity and its effect is measured in space-time. Imagine where you are right now? Where were you last night and where you will be at tonight? See time and space goes together. So your position and time together express an events. Such as last night you went to Jasper to see Rocky Mountain, or say tomorrow you will go to Toronto to attend a birthday party. You can't be more than one place at the same time. Each define time and place (space) together define or represent an events. Similarly in space time continuum each point define an events. Time is another dimension with three dimensions of space. If you learn to move or travel at this dimension then you can go back to past or travel to the future. It is possible only at or near the speed of light. We know nothing can go faster than the speed of light in the universe. Time and space get bend, time and space is stretched or contracted to keep the light speed constant. The interesting thing is that if you travel in the space your clock will slow down, you will win the time. This is Albert Einstein's theory of relativity that has been tested by high-speed aircraft. The speed limit of the universe is the speed of light. Nothing can go faster than light. If you can travel in space with the speed of light or close the speed of light then you will be stuck with time, you will not grow old as you suppose to grow in earth. When you will be back in earth you will see your twin sibling may be 99 years old and telling a story about your space adventure to her grandchildren.

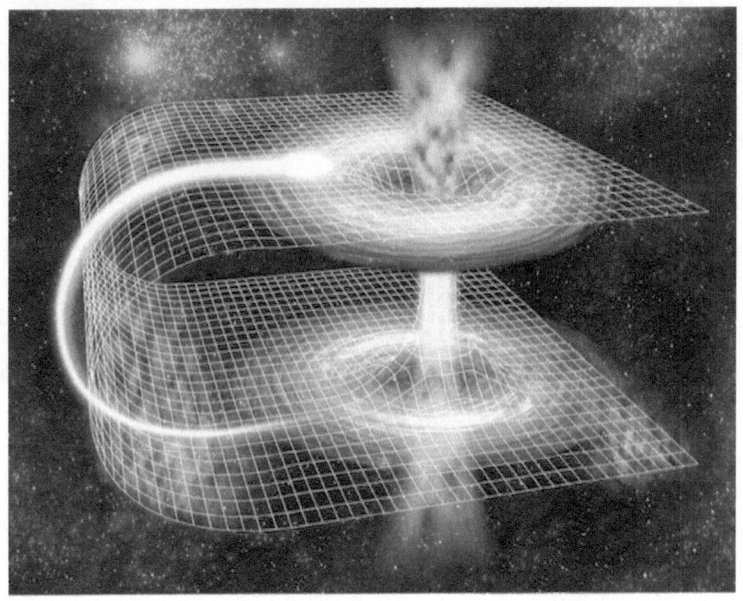

Source: Space.com

Wormhole is the place where the twisted or wrapped place in space-time continuum. It is a shortcut road between the distance space. Wormhole may be the connecting tunnel between Black Hole and imaginary White Hole. The black hole swallows everything, even drag the light into it. In reality it exists. For example, there is a huge black hole in the center of our galaxy. And white holes are imaginary thing which emit light. Opposite to the black hole, where something from the outside cannot be entered. An imaginary area of space-time continuum.

In 1905 Albert Einstein invent his famous $E = mc^2$ formula. This demonstrates mass (m) and energy (E) the same thing but only different conditions. This equation states that a large amount of energy is hidden in the small mass or matter. For starters, the **E** stands for **energy** and the **m** stands for **mass**, a measurement of the quantity of matter. Energy and matter are interchangeable. Furthermore, it's essential to remember that there's a set amount of energy/matter in the universe. Albert Einstein's most famous equation shows that mass can be converted to energy, and energy can be converted to mass. This means, in essence, that mass and energy are equivalent concepts. The energy produced by complete conversion of mass to energy is equal to the mass of an object times the speed of light squared. Note that this formula applies to the "rest mass" of an object. For fast-moving objects, special relativity applies, and a different formula is required to find the total energy.

## Can the child born in the space live in the earth?

In space where there is no gravity, if a child is born and grows there, he will not be able to live in the earth. Growing up in a zero gravity environment and growing in the influence of gravity are not the same. Our body is growing in the influence of Earth gravity and gravity has an important contribution in shaping our body system. Astronauts lose their muscle and bone faster and the heart becomes weak. Human beings growing in micro gravity will have very small and weak heart because heart does not require much energy to circulate blood. As a result, when they will come to earth, their heart will not have enough power to circulate blood against the earth gravity. For the influence of a gravitational force our blood wants to go down to feet, but our heart pump the blood in opposite direction of gravitational force. The heart circulate blood from feet to head and all over the body. For this, reason our heart is very strong in earth. But heart does not require that much power in the space as there is no gravity, so astronaut's legs get slim and their heart becomes weaker. To avoid this they had to do few hours of exercise every day. On the other hand if we go in Jupiter, our heart will not be able to circulate blood through our body as Jupiter has very high gravity compare to earth.

## What is gravity?

The gravity keeps us on the earth's surface, it keeps the planets orbiting around the sun. The gravitational force field is spreads around at the speed of light. This weak force compare to strong nuclear force and electromagnetic force but its field is very wide. Think our moon is quarter million miles away and still under earths gravitational field. If sun did not have the gravity, we would go out of the solar system. A space station creates a very small gravity known as micro-gravity that holds an astronaut.

Canadian Astronaut Chris Hadfield at International space station

Anything that has mass also has gravity. Objects with more mass have more gravity. Gravity also gets weaker with distance. So, the closer objects are to each other, the stronger their gravitational pull is. Earth's gravity comes from all its mass. All its mass makes a combined gravitational pull on all the mass in your body. That's what gives you weight. And if you were on a planet with less mass than Earth, you would weigh less than you do here. Gravity not only pulls on mass but also on light. Albert Einstein discovered this principle. If you shine a flashlight upwards, the light will grow imperceptibly redder as gravity pulls it. You can't see the change with your eyes, but scientists can measure it. Recently gravitational wave was detected by scientist which was predicted by Einstein about 100 years back.

## What is black hole?

Black holes are the strangest objects in the Universe. A black hole does not have a surface, like a planet or star. Instead, it is a region of space where matter has collapsed into itself. This catastrophic collapse results in a huge amount of mass being concentrated in an incredibly small area. The gravitational pull of this region is so great that nothing can escape – not even light. Although black holes cannot be seen, we know they exist from the way they affect nearby dust, stars and galaxies. (Ref ESA)

A black hole is a place where laws of physics does not work. Black hole is not an actual hole in space, rather it is a singular point that absorb everything, even light. A hole through a reality fabric of curved Space time continuum. If you grip a sheet of fabric surround it, a field will be created by the fabric. Now if you put a heavy ball in it, you will see that the fabric has been twisted downward. The heavier the ball is, the more it will bend the fabric. In space time continuum the thread of the fabric is space and time. The curvature effect is the gravity. When a star burns through the last of its fuel, it may find itself collapsing. For smaller stars, up to about three times the sun's mass, the new core will be a neutron star or a white dwarf. But when a larger star collapses, it continues to fall in on itself to create a stellar black hole. Black holes formed by the collapse of big stars into a tiny area, it is small but incredibly dense. Such an object packs three times or more the mass of the sun packed into a city-size range. This leads to a crazy amount of gravitational force pulling on objects around it. Black holes consume the dust and gas from the galaxy around them, growing in size.

Escape velocity is the speed that an object needs to escape of a planet or moon's gravity well and leave it without further propulsion. For example, a spacecraft leaving the surface of Earth needs to be going 7 miles per second, or nearly 25,000 miles per hour to leave without falling back to the surface or falling into orbit. The escape velocity from the surface (i.e., the event horizon) of a black hole is exactly same as the speed of light or higher. The escape velocity from inside the black hole is higher than the speed of light but we know the speed limit of the universe is the speed of light. Nothing can go faster than the speed of light.

Small stars do not create black holes: When smaller stars "Die" they become White Dwarfs. It takes a massive star to create Supernova must be 10-15 times more massive than our Sun. After this Supernova, if there is enough remaining debris nearby it can be pulled together by gravity and create either a neutron star or black hole. If the conditions are right, the debris will be pulled into a single point, creating a singularity with infinite density. You need an imagination to realize the mysterious universe.The formation and evolution of stars and is so lengthy, the distance is so big, and the temperature is so high that it is impossible to experience this extreme observation with our senses. This can be unfamiliar to us. Can we really feel the power of our sun, the size of orbit, supernova explosion and its brightness which enlighten almost the whole galaxy?

## What happens when we go to black hole?

Black hole tension is so strong that it will pull you up and stretch you from head to toe. It influence the time. Your fiends will see you as similar to seeing through magnifying glass. As you approach close to your friend will see your speed is slow down. If you call your fiend using your flash light then its speed will slow. When you go at the event horizon then your friend will see your time has stopped. Your time will stop as like paused movie on TV screen. As soon as you pass the event horizon your friend will see you are disappear into dark and you will see everything faster. Outside the black time travel only in one direction that is only forward, it never goes backward. But inside the black hole the space time bend such a way that the exchange each other. If you like you may go back in time. Inside the black hole the laws of physics collapse.

## How Star is born and die?

About four and a half billion years from today the sun will end up with its fuel. Then it will turn into a red giant, it will swell so big that it swallows Mercury and Venus. And the earth will then become the closest planets of the sun. And the heat of the sun will burn the earth surface. But outer colder layer will become worm. If the human species survive those days, maybe there will be move to other planet or star system. When a medium-sized star power dissipates, it turn into a white dwarf. The gravitational force in the center of the star tends to pull the mass of the star inwards, and it collapses inside. Our sun will follow the same route.

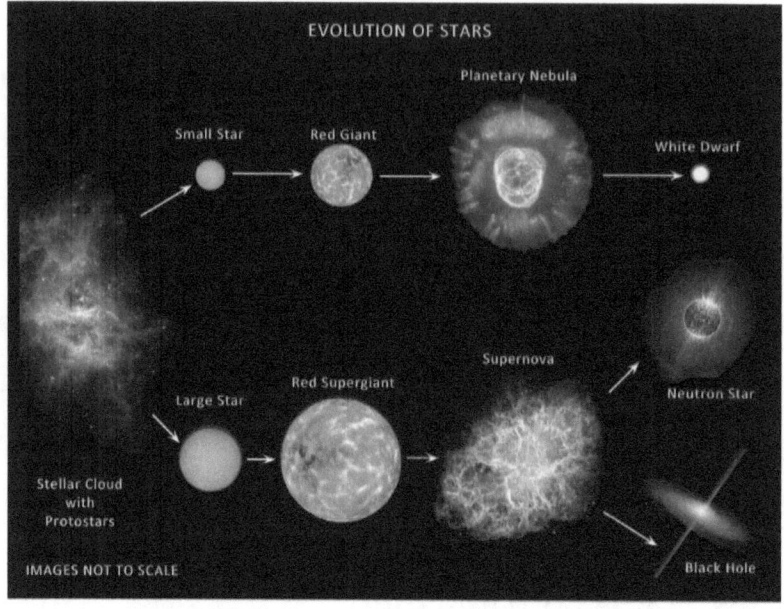

Steller Formation system (Image: NASA)

When a massive star runs out of fuel, it cools off. This causes the pressure to drop. Gravity wins out, and the star suddenly collapses. Imagine something one million times the mass of Earth collapsing in 15 seconds! The collapse happens so quickly that it creates enormous shock waves that cause the outer part of the star to explode! Usually a very dense core is left behind, along with an expanding cloud of hot gas called a nebula. A supernova of a star more than about 10 times the size of our sun may leave behind the most dense objects in the universe—black holes.

A second type of supernova can happen in systems where two stars orbit one another and at least one of those stars is an Earth-sized white dwarf. A white dwarf is what's left after a star the size of our sun has run out of fuel. If one white dwarf collides with another or pulls too much matter from its nearby star, the white dwarf can explode.

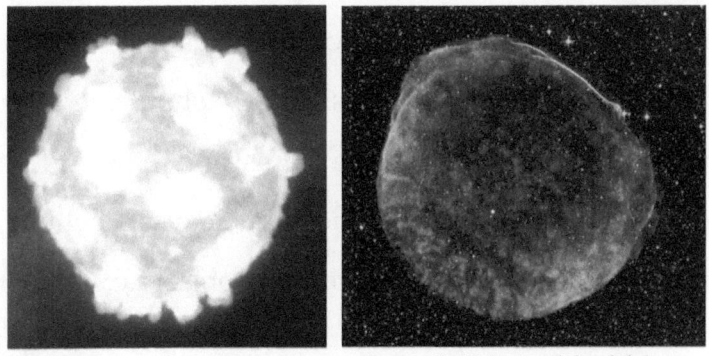

Supernova explosion capture by ANSA and bubble shaped remnant after supernova (Image: NASA)

## What is Nebulae?

The interstellar gas consists partly of neutral atoms and molecules, as well as charged particles (plasma), such as ions and electrons. This gas is extremely dilute, even though the interstellar gas is much dispersed, the amount of matter adds up over the vast distances between the stars. And eventually, and with enough gravitational attraction between clouds, this matter can coalesce and collapse to forms stars and planetary systems.

Mutual gravitational attraction causes matter to clump together, forming regions of greater and greater density. From this, stars may form in the center of the collapsing material. When the stars dies through supernova explosion then its remnant is spread across the universe. Some nebulae are formed as the result of supernova explosions, and are hence classified as a Supernova Remnant Nebulae. In this case, short-lived stars experience implosion in their cores and blow off their external layers. This explosion leaves behind a "remnant" in the form of a compact object i.e. a neutron star and a cloud of gas and dust that is ionized by the energy of the explosion. Observations suggest the star ejected its mass in a series of pulses at 1,500-year intervals. These convulsions created dust shells, each of which contain as much mass as all of the planets in our solar system combined (still only one percent of the Sun's mass). These concentric shells make a layered, onion-skin structure around the dying star. The view from Hubble is like seeing an onion cut in half, where each skin layer is visible.

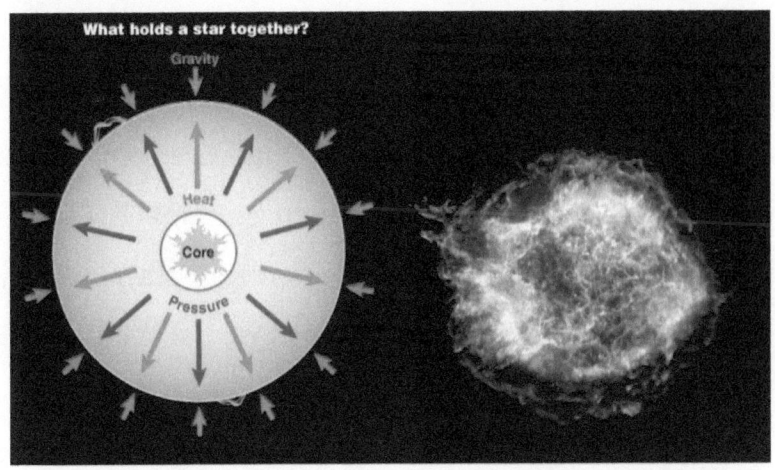

Gravity and pressure balance the star to burn, once it loose the balance it dies and the remnant spread around the universe

**The Cat's Eye Nebula** (NGC 6543) is revealed from NASA's Hubble Space Telescope. The image from Hubble's Advanced Camera for Surveys (ACS) shows a bull's eye pattern of eleven or even more concentric rings, or shells, around the Cat's Eye. Each 'ring' is actually the edge of a spherical bubble seen projected onto the sky - that's why it appears bright along its outer edge.

**NGC 7662** is a blue snowball. Maybe one day our Sun will look like this. The Blue Snowball is a planetary nebula - and in 5 billion years the Sun will throw off its outer layers and go through a planetary nebula phase. A star can appear "normal" only so long as there are sufficient nuclear reactions in its core. Soon thereafter, gravity will win out and compress the stellar core to higher temperatures. Eventually the core becomes a white dwarf. These high temperatures somehow cause the expulsion of star's outer layers, creating a planetary nebula such as the Blue Snowball pictured above.

The layers of the **NGC 7009** Nebula give a complex picture of how this planetary nebula was formed. This allows us a better understanding of the mysterious process that transformed a low-mass star into a white dwarf star. A computer model indicates that the central star of **NGC 7009** first expelled the green gas that now appears barrel shaped. This green gas now confines stellar winds flowing from the central star, creating a jet.

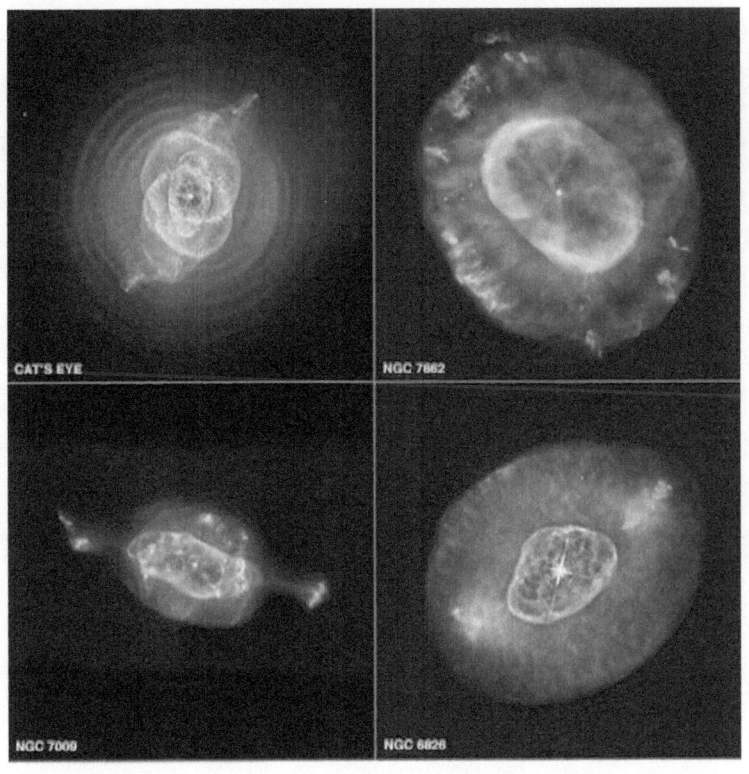

The colorful planetary nebula phase of a sun-like star's life is brief. Almost in the "blink of an eye" - cosmically speaking - the star's outer layers are cast off, forming an expanding emission nebula. This nebula lasts perhaps 10 thousand years compared to a 10 billion year stellar life span. Spectacular planetary nebulae are familiar objects to both professional and amateur astronomers, but they still contain a few surprises. For instance, the lovely nebula **NGC 6826**, also known as the Blinking Eye Nebula, has mysterious red FLIERS seen on either side of the Hubble Space Telescope image above. Are they also expanding outward from the central star? If so, their "bow shocks" point in the wrong direction! (Ref: NASA)

## We are made out of star dust ?

Do you know that if we die, new stars will be born again from you? You know that your body is mainly compose of hydrogen, oxygen and carbon. This is because more than half your body weight is water. All the life on earth is mostly compose of hydrogen, oxygen, carbon and the rest of the other components. The hydrogen of your body was formed during the birth of the universe. The rest of the elements that make you up was formed into the stars and they just recycled repeatedly. When life dies, its body get decompose into elements and again those elements are mixed together to build something new. It's just as you build a thing with the smallest piece of Lego (Lego) and break them and make new things with the same Lego pieces. The process used in nature is chemical process. The elements used to build air, soil, water cloud, the earth, solar system will be used another star system. Our solar system also was form the remnant of a dyeing star. Who knows where my element was used and it might be used to form anther intelligent life in future. We are all connected, we are connected to everything, the whole cosmos.

## How our solar system born and evolve?

How can we make a theory of something that happened 4.6 billion years ago? We didn't observe the origin of the Solar System, so we have to develop theories that match "circumstantial evidence" - what the Solar System is like today. The star formation theories, observation, the patterns we see today, Processes of planet formation are used to explain the theory. The rock from meteor, moon, and earth has been analyzed by geologiest to collect the evidence. About 4.6 billion years ago, the Solar System was formed by gravitational collapse. The solar system coalesced from a huge cloud of dust and gas that was isolated from the rest of the Milky Way galaxy. This primary molecular cloud was broadly expanded for a few light years, and it is thought that several stars were created from it. Modern research with meteorites has found some rare elements that can only be produced within an explosive star. From this it is understood that the Sun was born in a star constellation and from a remnant of a massive supernova explosion.

Initially the cloud was about several light years across. A small over density in the cloud caused the contraction to begin and the over density to grow, thus producing a faster contraction. Initially, most of the motions of the cloud particles were random, yet the nebula had a net rotation. As collapse proceeded, the rotation speed of the cloud was gradually increasing due to conservation of angular momentum. Gravitational collapse was much more efficient along the spin axis, so the rotating ball collapsed into thin disk with a diameter of 200 AU (0.003 light years), with most of the mass concentrated near the center. As the cloud contracted, its gravitational potential energy was converted into kinetic energy of the individual gas particles. Collisions between particles converted this energy into heat (random motions). The solar nebula became hottest near the center where much of the mass was collected to form the proto-sun (the cloud of gas that became Sun). At some point the central temperature rose to 10 million K. The collisions among the atoms were so violent that nuclear reactions began, at which point the Sun was born as a star, containing 99.8% of the total mass.

What prevented further collapse? As the temperature and density increased toward the center, so did the pressure causing a net force pointing outward. The Sun reached a balance between the gravitational force and the internal pressure, aka as hydrostatic equilibrium, after 50 million years. Around the Sun a thin disk gives birth to the planets, moons, asteroids and comets. Over recent years we have gathered evidence in support of this theory. Close-up of the Orion Nebula obtained with HST, revealing what seem to be disks of dust and gas surrounding newly formed stars. These protoplanetary disks span about 0.14 light years and are probably similar to the Solar Nebula.

Orion Nebulae (Image: NASA)

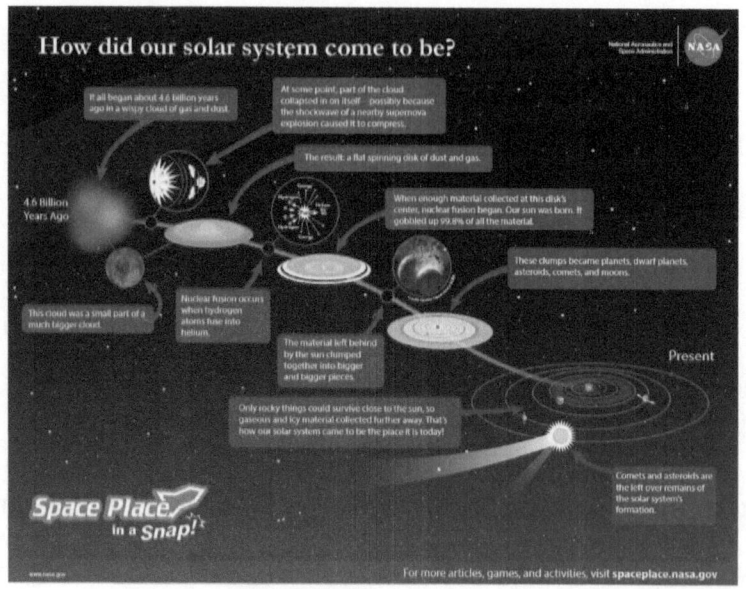

NASA's Spitzer and Hubble Space Telescopes teamed up to expose the chaos inside the cosmic cloud known as Orion nebulae. Scientist and Astronomer found how the baby stars are forming around 1,500 light years away from us. This striking composite indicates that four monstrously massive stars, collectively called the "Trapezium," at the center of the cloud. The Orion constellation, a familiar sight in the fall and winter night sky in the northern hemisphere.

**What is solar system?**

What do we know about our sun and solar family? Our sun is a medium size star of millions of stars in our own Milky Way galaxy. The source of all our energy is the sun. The sun is important for us because it gives us heat and light energy. All life in the earth are dependent on the sun. All the solar planets including our earth and the sun are all floating in space. All these things are in accordance with the rules that say solar system or solar system. The Sun contains 99.85% of the mass. The Solar System is mostly empty space. The Solar System is a flattened disk. All planets revolve around sun in the same direction and most planets also rotate on own axis in the same direction. All objects have same ages about 4.6 billion years. All the orbits are all in a single plane. The Sun's equator lies in this plane. Planetary orbits are nearly elliptical.

**What we have in our solar systems?**

The Solar System refers to all the astronomical objects which are bound by the sun and its gravitational effects. There are seven planets (Mercury, Venus, Earth, Mars, Jupiter, Saturn, Uranus, and Neptune). The satellites of the planets; numerous comets, asteroids, and meteoroids; and the interplanetary medium. The Sun is the richest source of electromagnetic energy (mostly in the form of heat and light) in the solar system. The Sun's nearest known stellar neighbor is a red dwarf star called Proxima

Centauri, at a distance of 4.3 light years away. The Sun contains 99.85% of all the matter in the Solar System. The planets, which condensed out of the same disk of material that formed the Sun, contain only 0.135% of the mass of the solar system. Jupiter contains more than twice the matter of all the other planets combined. Satellites of the planets, comets, asteroids, meteoroids, and the interplanetary medium constitute the remaining 0.015%.

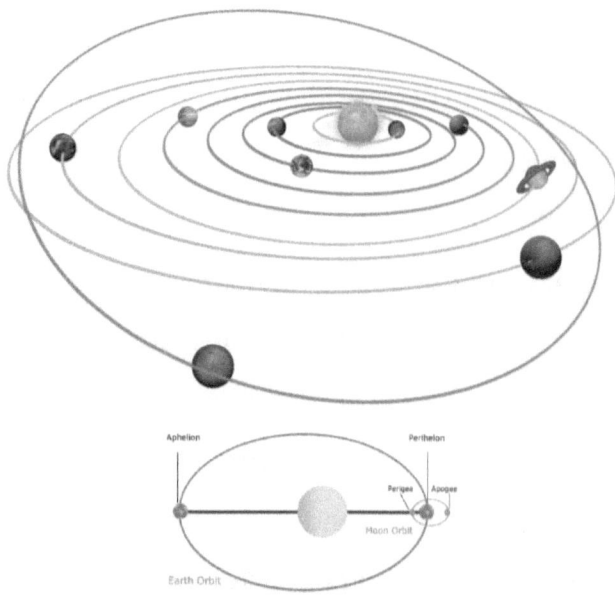

**Orbit:** An orbit is a regular, repeating path that one object in space takes around another one. An object in an orbit is called a satellite. A satellite can be natural, like

Earth or the moon. Many planets have moons that orbit them. A satellite can also be man-made, like the International Space Station. Planets, comets, asteroids and other objects in the solar system orbit the sun. Most of the objects orbiting the sun move along or close to an imaginary flat surface. This imaginary surface is called the ecliptic plane. (Ref: NASA)

**Inner solar system**

The four inner four planets Mercury, Venus, Earth and Mars are made up mostly of iron and rock. They are known as terrestrial or earth like planets because of their similar size and composition. Earth has one natural satellite, the moon and Mars has two moons known as Deimos and Phobos. Between Mars and Jupiter lies the Asteroid Belt. Asteroids are minor planets, and scientists estimate there are more than 750,000 of them with diameters larger than three-fifths of a mile (1 km) and millions of smaller asteroids. The dwarf planet Ceres, about 590 miles (950 km) in diameter. A number of asteroids have orbits that take them closer into the solar system that sometimes lead them to collide with Earth or the other inner planets.

**Outer solar system**

The outer planets – Jupiter, Saturn, Uranus and Neptune are giant worlds with thick outer layers of gas. These planets, they have dozens of moons with a variety of compositions, ranging from rocky to icy. Some even has volcanic action such as Jupiter's moon Lo. Nearly all the planets' mass is made up of hydrogen and helium, giving them compositions like that of the sun. Beneath these outer layers, they have no solid surfaces – the pressure from their thick atmospheres liquefy their insides, although they might have rocky cores. Rings of dust, rock, and ice encircle all these giants, with Saturn's being the most famous.

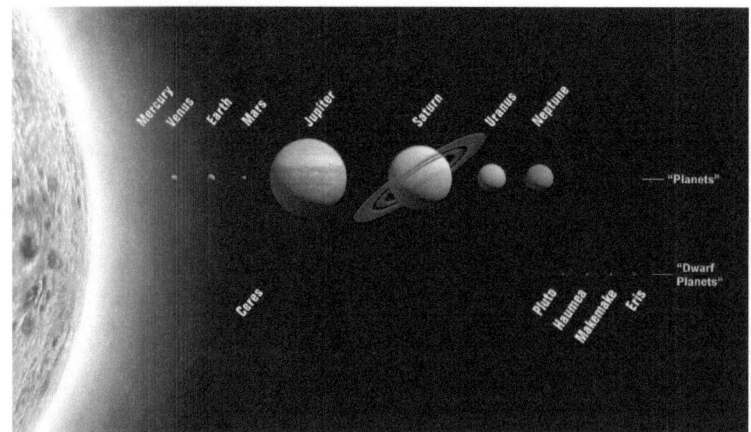

**Trans-Neptunian region**

A band of icy material known as the Kuiper Belt exist past the orbit of Neptune extending from about 30 to 55 times the distance of Earth to the sun. The Kuiper Belt is likely home to hundreds of thousands of icy bodies larger than 60 miles (100 km) wide, as well as an estimated trillion or more comets. Pluto, now considered a dwarf planet, dwells in the Kuiper Belt.

It is not alone, it include Make-make, Haumea and Eris. Another Kuiper Belt object dubbed Quaoar is probably massive enough to be considered a dwarf planet, but it has not been classified as such yet. The Oort cloud lies well past the Kuiper Belt, and theoretically extends from 5,000 to 100,000 times the distance of Earth to the sun, and is home to up to 2 trillion icy bodies, according to NASA. Past the Oort cloud is the very edge of the solar system, the heliosphere, a vast, teardrop-shaped region of space containing electrically charged particles given off by the sun. Many astronomers think that the limit of the heliosphere, known as the heliopause, is about 9 billion miles (15 billion km) from the sun.

The sun is the main component of the solar system. 99.85 percent of the total mass of the solar system is the sun, and it controls everything. Jupiter and Saturn are responsible for the most percent of the remaining mass of the solar system, except the sun. These planets are the largest objects orbiting in the sun. Most of the objects orbiting around the center of the sun are located on the same level. The planets are located on this same plane, although comets and other Kuiper belt objects are orbit in an angle. The planets and most other objects are rotating in the opposite direction of the clock. Also some exception such as Halley Comet the objects orbiting around the sun obey Kepler's planetary motion. Each object rotates in the elliptical orbit by keeping the sun at a focus on the ellipse. The speed of the object increases as it approaches the sun.

**Asteroid Belt**

About 2-4 AU (186-370 million miles) away from the Sun, between the orbits of Mars and Jupiter, is a region called the Asteroid Belt. This region is a ring of tens of thousands of relatively small rocky objects called Asteroids. Scientist assumed that during the formation of other planets, another planet was about to form between Mars and Jupiter .but due to low mass, gravity and

Jupiter's gravity prevented the material in the belt from coalescing into larger planets. During the formation of solar system, due to gravitational attraction of the planet Jupiter, the objects that could not be formed together in large objects were trapped in this enclosure. The largest asteroid discovered so far is the Ceres. Its diameter is 945 kilometers. Asteroids are made up of metallic rocky minerals. The asteroids are not embedded in a very dense form. Spacecraft from the Earth could cross this belt easily without any collision.

**Kuiper Belt**

The Kuiper Belt is a disc-shaped region of icy bodies - including dwarf planets such as Pluto - and comets beyond the orbit of Neptune. It extends from about 30 to 55 AU and is probably populated with hundreds of thousands of icy bodies larger than 100 km (62 miles) across and an estimated trillion or more comets. The first Kuiper Belt Object was discovered in 1992 by astronomer Gerard Kuiper. (NASA). Astronomers are now hunting for more object in the Kuiper Belt. When the solar system formed, much of the gas, dust and rocks pulled together to form the sun and planets. The planets then swept most of the remaining debris into the sun or out of the solar system. But bodies farther out remained safe from gravitational tugs of planets like Jupiter, and so managed to stay safe as they slowly orbited the sun. The Kuiper Belt and its compatriot, the more distant and spherical Oort cloud, contain the leftover remnants from the beginning of the solar system and can provide valuable insights into its birth.

**The Sun**

There are so many variations of life in our beautiful planet earth. Earths all energy sources is the sun. There are so much diversity in life in earth because sun continuously sending the energy. Life cannot be imagined in any way without the sun light. Have you ever wondered how the light is created in the sun, and how does this light come to earth? The thermonuclear fusion is a process by which light, heat, and energy are produced in a star. It occurs in the center of the star, the temperature in the center is up to millions degrees. This heat travels towards the surface of the star and radiates to the universe. Hydrogen is used as fuel in this process. This resulted in the formation of another type of gas called Helium. The fire of star and the fire that we use in our earth is not the same, it does not burn in the same way. In earth we burn something to make fire is the chemical process, the material produces heat and light in the chemical reaction with oxygen from the air. Star fire is like a nuclear power plant or nuclear bomb.

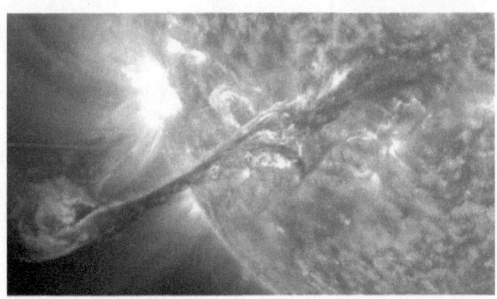

The sun's distance from Earth is 150 million kilometers. And the speed of light is 300,000 kilometers per second. It takes 8 minutes to come from Sun to Earth .Earth's distance from sun is called 1 astronomical unit or AU .The Sun creates energy from the nuclei fusion from its hydrogen gas at its center. This power comes from the center of the surface and spreads around in the form of light and heat. This energy takes about 100,000 years to come to the surface from the Sun's center.

**How the sun makes energy?**
**Nuclear Fusion:** We know light is the photon particle which travels as wave. The light that comes from the sun is caused by the nuclear fusion. Photon particles are formed due to nuclear fusion and they are spread as

waves of light. To understand nuclear fusion it required to know about the structure of atoms. The atomic center is called nucleus. Proton and neutron stays together in the nucleus. Around the nucleus, the electrons orbits in their own orbits, stays like a cloud. Proton has positive charge, neutron no charge and electron has negative charge. The force that contains protons and neutrons in the nuclei together known as strong nuclear force. This is one of the force that governs the universe. All the objects in the universe are made by holding like a glue by this force or energy. This resulted in the whole universe being enlightened. If this doesn't exist, there would be no atom, no star, no planet, and no molecule. There would no creation of the earth, no life and we would not exist. The whole universe would be kept in darkness. We have learned how to use this energy through nuclear power plant that use nuclear fusion process to create energy. The process of nuclear fusion is very complex and its results are wide. Especially in the universe it is so vast that it cannot be easily imagined. About 75 percent of the sun's mass is hydrogen gas. Hydrogen atoms in the Sun create fusion (meaning, paired) continuously and form relatively large atom of helium and energy release from these process. In find the infinite mystery of universe, a scientific mind will be created inside you. Your curiosity will reveal the mystery of the universe. Another name for the Sun's nuclear fusion is proton-proton chain reduction, which allows the protons of two different atoms to fuse and form a new mass of atoms. We know that protons are usually positive charged, then how they combine two protons? But remember that 99.8 percent of the total mass our solar system is belongs to the Sun. Due to this massive mass, the gravity of the Sun is incredible high, which is about 24 times more than the Earth. Due to this deadly gravitational force the surrounding gas pressure at center is very high. This pressure cause the temperature in the center of the sun very very high. This high energy force the hydrogen gas in the center of the sun to add or fuse or to combined together to form helium. This high temperature and pressure force two repulsive proton to be fused or to be added. And release photon particle through the radiation.

This process is happening continuously at the very center of the sun which is called the core. But after the formation of the light particle, it cannot directly reach the sun surface. It has to cross two more zone, which is the radiation and conduction zone. In these radiation zone the light particles are constantly interrupted, are absorbed, are again scattered. Then it comes through the sun's conduction surface zones. Since the formation of the particle to come to surface, the whole process to come to the surface takes 10,000 to 170,000 years. It means that the light you get now is thousands of years old, it was formed thousands of years ago in the center of the Sun.

**What is the structure of sun?**
From the far we see the sun as a huge yellow steady sphere. But inside and outside the Sun is not at all still. We have learned how the nuclear fusion is happening at the center. And it is blowing energy from its surface flooding the entire solar system with light. The surface of the sun is not as strong as our earth, it is a burning hot ball. Due to high heat and pressure, the substance (matter), which is mainly produced by a hydrogen gas is in form of plasma. The boiling hot plasma sphere does not stay same every day. The sun constantly bursting through the explosion. The sun is not just spreading heat and light, it is fluttering charged particles at every moment. The flames are flowing at every moment. This violent video can be seen in NASA website. Scientist are trying to understand the sun last few years but Just for

the last 20 years we have seen the sun from very close to the terrestrial satellite with special telescope.

**Inner core:** This is where fusion occurs. The core is the very center of the sun where fusion occurs. The core has extremely high temperature and pressure. It is so hot here around 15 million degree Celsius. At this temperature nuclear fusion triggers. In the case of our sun, four hydrogen nuclei are fused into one helium nuclei while releasing a bunch of energy as photons. The core is the hottest part of the sun and the only part of the sun that produces much energy. The sun cools off as you travel from the core to the outside, with the exception of the chromosphere.

**Radiative zone:** The layer of the sun directly above the core. This zone, as you can probably guess, emits radiation, and the radiation from the core diffuses out from here. It may take photons millions of years to get out.

**Convective zone:** In this layer, photons produced by fusion in the core make their way to the surface of the sun through convection. This zone is dominated by convection currents that carry the energy produced through fusion in the sun's core outward to the surface. A convection current is when hot gas rises next to hot gas falling, which causes movement or currents. This energy is carried by photons which can take 170,000 years to move from the core to the sun's surface. It then takes eight minutes for those photons to travel to Earth and provide us with energy.

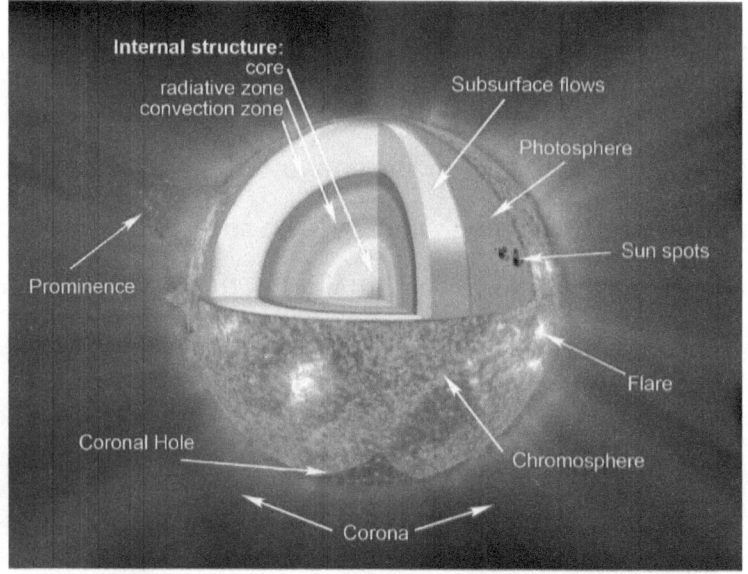

Structure of our Sun (image:NASA)

**Photosphere:** The first of the outer layers of the sun is the photosphere. This is the part of the sun that we see because it emits light in our visible wavelengths. This layer is about 300 miles thick, and the temperature has cooled to 5,500 degrees Celsius. There is not really a 'surface' to the Sun, but this is the layer that we see that looks like it is the surface.

**Chromosphere:** The chromosphere is a layer in the Sun between about 250 miles (400 km) and 1300 miles (2100 km) above the solar surface (the photosphere). The temperature in the chromosphere varies between about 4000 K at the bottom (the so-called temperature minimum) and 8000 K at the top (6700 and 14,000 degrees F, 3700 and 7700 degrees C), so in this layer (and higher layers) it actually gets hotter if you go further away from the Sun, unlike in the lower layers, where it gets hotter if you go closer to the center of the Sun.

**Transition Region:** The transition region is a very narrow (60 miles / 100 km) layer between the chromosphere and the corona where the temperature rises abruptly from about 8000 to about 500,000 K (14,000 to 900,000 degrees F, 7700 to 500,000 degrees C) The sun burns 4 million tons of matter every second to produce energy. Yes, the sun will eventually burn out. The sun has used up about half of its hydrogen fuel in the last **4.6 billion years**, since its birth. There is nothing to be afraid of, the amount of fuel sun has will keep burning for another 5 billion years.

**Corona**: The corona is the outermost layer of the Sun, starting at about 1300 miles (2100 km) above the solar surface (the photosphere). The temperature in the corona is 500,000 K (900,000 degrees F, 500,000 degrees C) or more, up to a few million K. The corona cannot be seen with the naked eye except during a total solar eclipse, or with the use of a coronagraph. The corona does not have an upper limit. Credit: National Solar Observatory (Refines)

**What is Aurora?**
When the activity of the sun rises, its enormous amounts of gas and plasma are released into the atmosphere. A solar flare occurs when magnetic energy that has built up in the solar atmosphere is suddenly released coronal mass ejections (CME). Like the sudden release of a twisted rubber band, the magnetic fields explosively realign, driving vast amounts of energy into space. This phenomenon can create a sudden flash of light. Flares can last minutes to hours and they contain tremendous amounts of energy. CMEs can funnel particles into near-

Earth space. A CME can jostle Earth's magnetic fields creating currents that drive particles down toward Earth's poles. When these react with oxygen and nitrogen, they help create the aurora, also known as the Northern and Southern Lights. The colors in the aurora were also a source of mystery throughout human history. But science says that different gases in Earth's atmosphere give off different colors when they are excited. Oxygen gives off the green color of the aurora, for example. Nitrogen causes blue or red colors. More details mechanism:

https://www.plasma-universe.com/Aurora

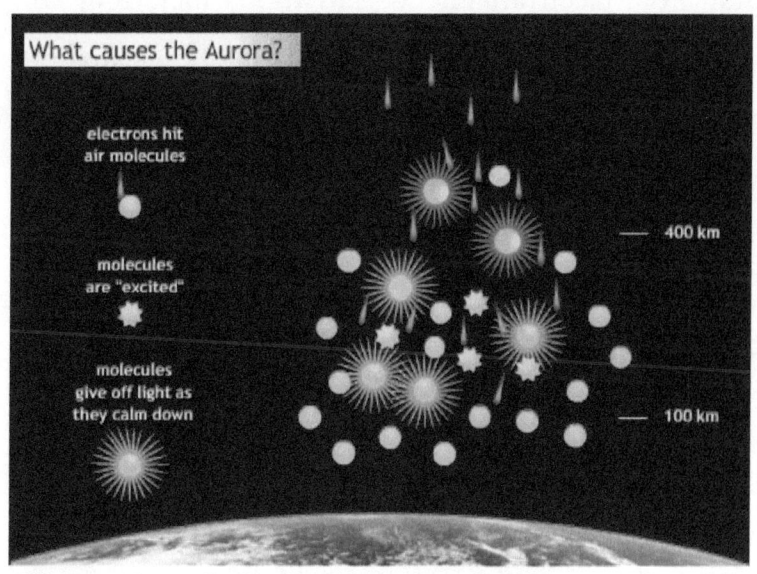

The northern light formation process (Image: NASA)

## What is solar flare?

When the activity of the sun rises, its enormous amounts of gas and plasma are released into the atmosphere. A solar flare occurs when magnetic energy that has built up in the solar atmosphere is suddenly released such as Solar flares and coronal mass ejections. Solar flares emit bursts of electromagnetic radiation including high energy X-rays and gamma rays. Solar flares are often followed by a large ejection of plasma from the surface of the Sun called a Coronal Mass Ejection (CME). Like the sudden release of a twisted rubber band, the magnetic fields explosively realign, driving vast amounts of energy into space. . This phenomenon can create a sudden flash of light. Flares can last minutes to hours and they contain tremendous amounts of energy. CMEs can funnel particles into near-Earth space. A CME can jostle Earth's magnetic fields creating currents that drive particles down toward Earth's poles. When these react with oxygen and nitrogen, they help create the aurora, also known as the Northern and Southern Lights. Flares and CMEs have different effects at Earth as well. The energy from a flare can disrupt the area of the atmosphere through which radio waves travel. This can lead to degradation and, at worst, temporary blackouts in navigation and communications signals.

**What is sunspots?**

Sunspots are dark areas that appear on the surface of the Sun. They are caused by strong magnetic activity within the sun. Sunspots are not permanent and they can move slowly across the surface of Sun changing size as they move. The appearance of sunspots follows the solar cycle of eleven years. Every eleven years there will be a period of increased sunspot activity. Sunspots vary in size from as small as 10,000 miles across to as large as 100,000 miles across. Sunspots are darker, cooler areas on the surface of the **sun** in a region called the photosphere. The photosphere has a temperature of 5,800 degrees Kelvin. Sunspots have temperatures of about 3,800 degrees K. They look dark only in comparison with the brighter and hotter regions of the photosphere around them.

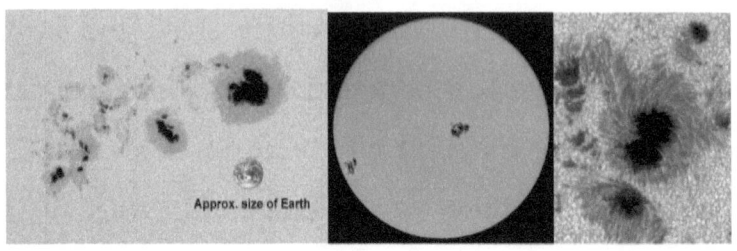

The sun spot can be seen by a telescope, this detail image taken by NASA

**What is solar wind?**

The sun burns hydrogen atom in its center, releasing vast amounts of atomic particles in space from the surface. This wind reaches beyond the solar system and mixes with other interstellar medium and creates a bubble around the sun, which is called as the heliosphere. This is considered as the boundary of solar system. The solar wind streams plasma and particles from the sun out into space. The solar wind streams off of the Sun in all directions at speeds of about 400 km/s (about 1 million miles per hour). The source of the solar wind is the Sun's hot corona. The temperature of the corona is so high that the Sun's gravity cannot hold on to it. Although we understand why this happens we do not understand the details about how and where the coronal gases are accelerated to these high velocities. This question is related to the question of coronal heating. The sun is blowing billions of kilometers solar wind outward, this thin stream of electrically charged particle travel until it reaches the termination shock point. At this point, the speed of the solar wind drops abruptly as it begins to feel the effects of interstellar wind .It creates a bubble that extends far past the orbits of the planets. This bubble is the heliosphere. The boundary between solar wind and interstellar wind is the heliopause, where the pressure of the two winds are in balance. This balance in pressure causes the solar wind to turn back and flow down the tail of the heliosphere. Once Voyager passes the heliopause, it will be in interstellar space. It shaped like a long wind sock as it moves with the Sun through interstellar space. As the heliosphere plows through interstellar space, a bow shock forms, much as forms in front of a boulder in a stream. (Ref: NASA)

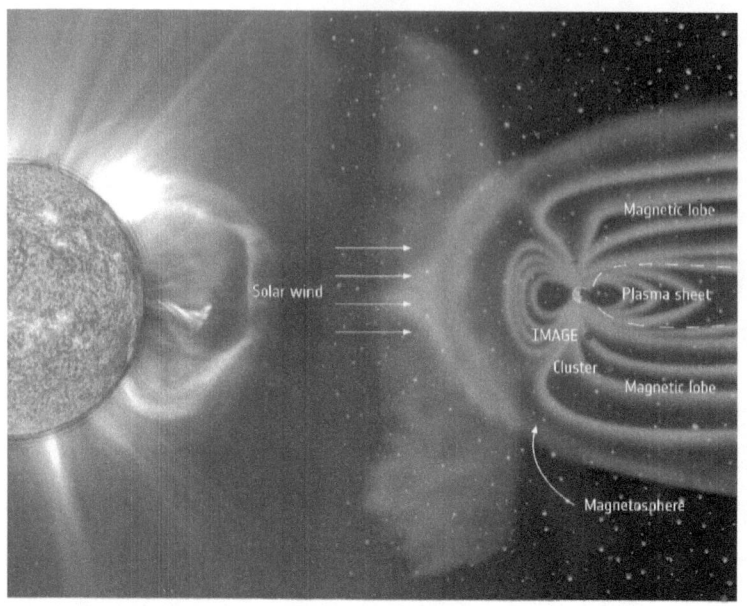

This ESA image shows how earth magnetic fields protect us from the dangerous solar wind

**What is a solar prominence?**

A solar prominence (also known as a filament when viewed against the solar disk) is a large, bright feature extending outward from the Sun's surface. Prominences are anchored to the Sun's surface in the photosphere, and extend outwards into the Sun's hot outer atmosphere, called the corona. A prominence forms over timescales of about a day, and stable prominences may persist in the corona for several months, looping hundreds of thousands of miles into space. Scientists are still researching how and why prominences are formed. The red-glowing looped material is plasma, a hot gas comprised of electrically charged hydrogen and helium. The prominence plasma flows along a tangled and twisted structure of magnetic fields generated by the sun's internal dynamo. An erupting prominence occurs when such a structure becomes unstable and bursts outward, releasing the plasma. (Ref: NASA)

The electromagnetic (EM) spectrum is the range of all types of EM radiation. Radiation is energy that travels and spreads out as it goes – the visible light that comes from a lamp in your house and the radio waves that come from a radio station are two types of electromagnetic radiation. The other types of EM radiation that make up the electromagnetic spectrum are microwaves, infrared light, ultraviolet light, X-rays and gamma-rays. The common term for electromagnetic radiation, usually referring to that portion visible to the human eye.

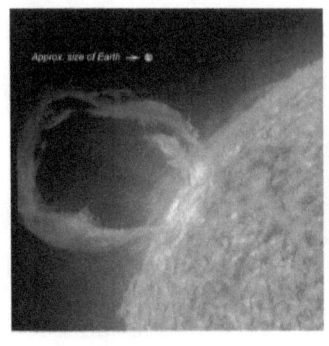 

## What is light?

Light is what helps us to understand our surroundings and the universe .Light falls in object from sources and reflect back to our eyes. Our eyes does have light receptor cell and it collect the information and send to our brain and finally our brain construct image based on light information we received. Light interact with matters so we recognize matters. Maxwell showed that electric and magnetic fields travel in the manner of waves, and that those waves move essentially at the speed of light. This allowed Maxwell to predict that light itself was carried by electromagnetic waves – which means light is a form of electromagnetic radiation. Light is a particle called photon and travel at speed of light?

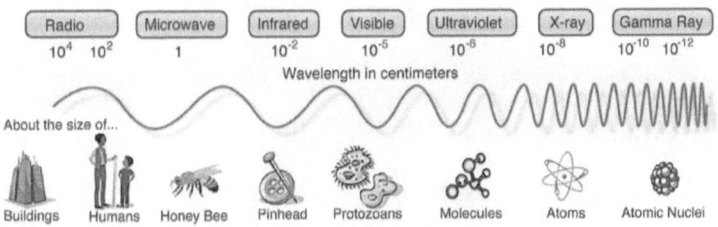

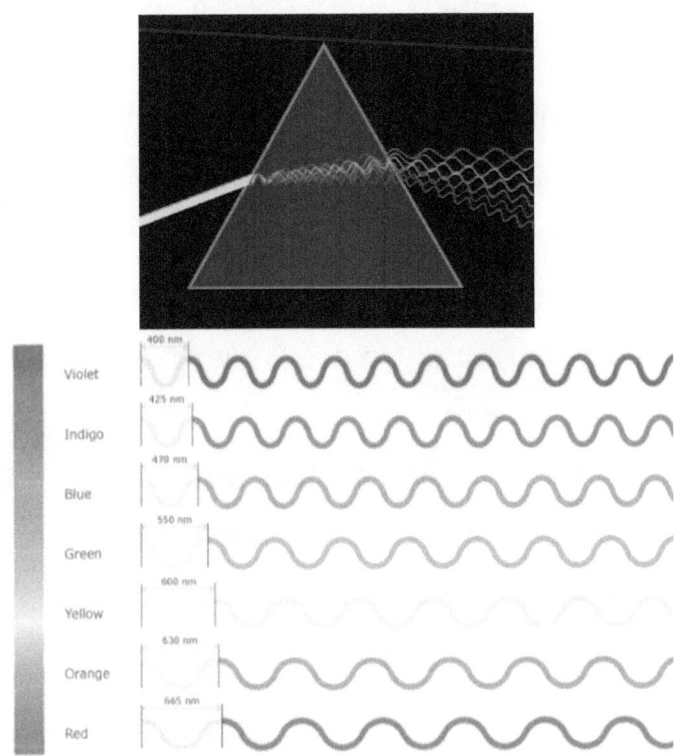

The universe was illuminated with light three hundred years after the birth through big bang. Light energy from the sun spreads in the space as photons. This photon is as particle and travel as wave. The full range of frequencies of the light is known as spectrum. A plot of the intensity of light at different frequencies. Or the distribution of wavelengths and frequencies is known as electromagnetic spectrum. However, other bands of the electromagnetic spectrum are also often referred to as different forms of light. But we only see a small portion of it known as visible spectrum. When you talk to your friend over the cell phone, do you see the signal? When you hear the music in radio, do you see the signal but both are a form of light. The electronic inside the radio or cell phone convert the light into sound. It was Newton who used the glass prism to prove first that light has seven colors of light. The wavelength and frequency of each color are very different. We only see the light of these seven waves only, this is known as viable spectrum, there NASA's scientific instrument unveils the mystery of the universe by studying the electromagnetic spectrum from different object in the universe. Radio: Your radio captures radio waves emitted by radio stations, bringing your favorite tunes. Radio waves are also emitted by stars and gases in space. Microwave: Microwave radiation will cook your popcorn in just a few minutes, but is also used by astronomers to learn about the structure of nearby galaxies. Infrared: Night vision goggles pick up the infrared light emitted by our skin and objects with heat. In space, infrared light helps us map the dust between stars. Visible: Our eyes detect visible light. Fireflies, light bulbs, and stars all emit visible light. Ultraviolet: Ultraviolet radiation is emitted by the Sun and are the reason skin tans and burns. "Hot" objects in space emit UV radiation as well. X-ray: A dentist uses X-rays to image your teeth, and airport security uses them to see through your bag. Hot gases in the Universe also emit X-rays. Gamma ray: Doctors use gamma-ray imaging to see inside your body. The biggest gamma-ray generator of all is the Universe.

**What is the interstellar medium?**

The empty space in space is not really empty. In between two star system inside the galaxy is called the interstellar medium (ISM). The most of the interstellar medium is made of clouds of hydrogen and helium gas and rest of them is carbon and other heavy elements. Voyager 1 and 2 space craft left the solar system boundary and travelling into the interstellar medium now. Still alive and sending signal to us. You can check its currents distance and location in NASA website. Space between the galaxy is called intergalactic medium (IGM). This is full of radiation and ionized gas.

The solar system moves through a local galactic cloud at a speed of 50,000 miles per hour, creating an interstellar wind of particles, some of which can travel all the way toward earth. When this wind collide with a solar wind then it create different layer Steller wind bubble. The solar wind flows outward from the Sun until encountering the termination shock with interstellar medium. The bubble-like region of space dominated by the Sun is known as **heliosphere.** The solar wind, creates and maintains this bubble against the outside pressure of the interstellar medium when orbiting the center of Milky Way galaxy. A transitional region which is in turn bounded by the outermost edge of the heliosphere, called the **heliopause**. The shape of the heliosphere is controlled by the interstellar medium through which it is traveling.

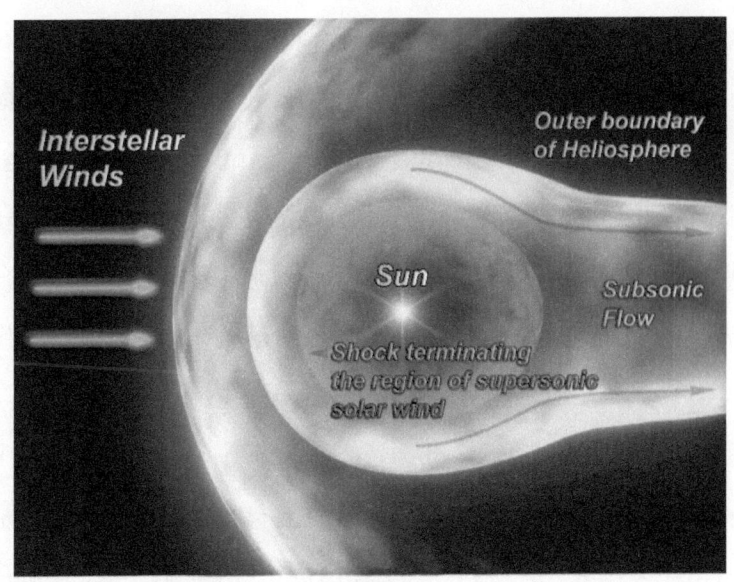

**What is interplanetary medium?**

The space between the planets is far from empty. It contains: electromagnetic radiation (photons); hot plasma (electrons, protons and other ions) a.k.a. the solar wind; cosmic rays; microscopic dust particles; and magnetic fields (primarily the Sun's). The supersonic outflow interacts with planetary magnetospheres and drives a number of magnetosphere phenomena, including geomagnetic storms on Earth.

**What is in the Solar system?**

The Solar System consists of planets, moons, comets, asteroids, minor planets, and dust and gas. Everything in the Solar System orbits or revolves around the Sun. The Sun contains around 98% of all the material in the Solar System. The larger an object is, the more gravity it has. Because the Sun is so large, its powerful gravity attracts all the other objects in the Solar System towards it. At the same time, these objects, which are moving very rapidly, try to fly away from the Sun, outward into the emptiness of outer space. The result of the planets trying to fly away, at the same time that the Sun is trying to pull them inward is that they become trapped half-way in between. Balanced between flying towards the Sun, and escaping into space, they spend eternity orbiting around their parent star. The planets in the solar system are divided into terrestrial and Jovian planets. They are different in their position, composition and other features. Mercury, Venus and Earth are the terrestrial planets. Jupiter, Saturn, Uranus and Neptune are the Jovian planets. The terrestrial planets are made of solid surfaces, the Jovian planets are made of gaseous surfaces. The Jovian planets are less dense when compared to the terrestrial planets, because they are mainly composed of hydrogen gas. Moreover, the core of the Jovian planets is denser than the terrestrial planets. The terrestrial planets are closer to the sun and the Jovian planets are farther. The Jovian planets are much larger than the terrestrial planets. While the atmosphere of terrestrial planets is composed mainly of carbon dioxide and nitrogen gases, hydrogen and helium gases are found in abundance in the atmosphere of Jovian planets. The Jovian planets have more moons than the terrestrial planets. Moreover, the Jovian planets tend to have rings around them, which are not seen in terrestrial planets. The terrestrial planets were much hotter when they were formed, and they cooled with time. The terrestrial planets were hit by meteorites during the early times, which made them so hot. This is why Earth and Venus have very hot interiors when compared to other planets.

**Mercury**

Mercury is the closest planet to the Sun, and the smallest planet of the solar system. Mercury has been visited by two unmanned NASA space probes, Mariner 10 and Messenger. This planet don't have any moon. Due to the high gravitation effect this closest planet to sun couldn't grow bigger during formation. It doesn't have atmosphere and solar wind effect is dangerous on this planet. Mercury's temperature is less than Venus, despite being close to the Sun. Since Mercury is the closest planet to the Sun, spins slowly, and does not have much of an atmosphere to trap heat, its temperature varies greatly. Mercury's temperatures can go between -279 Fahrenheit (-173 Celsius) at night to 801 Fahrenheit (427 Celsius) during the day.

**Venus**

Venus is a very similar size to the Earth, and like Earth, is made of a thick silicate mantle around an iron core. It has a substantial atmosphere and evidence of internal geological activity. Venus is probably the planet that is most similar to Earth in many ways, although it is much drier than Earth and its atmosphere is ninety times as dense. It is the hottest planet in the solar system, with surface temperatures over 400 °C. This is thought to be because of the amount of greenhouse gases in its atmosphere. The atmosphere of Venus is made up almost completely of carbon dioxide, with traces of nitrogen. Much of the hydrogen in the atmosphere evaporated early in the formation of Venus, leaving a thick atmosphere across the planet. At the surface, the atmosphere presses down as hard as water 3,000 feet beneath Earth's ocean. And thick clouds of sulfuric acid completely cover the planet.

**Earth**

Earth is the largest and densest of the four inner planets, the only one known to have current geological activity, like earthquakes and volcanoes. It is the only planet known to have life. Its liquid hydrosphere (oceans and seas) is unique among the terrestrial planets. Earth's atmosphere is radically different from those of the other planets, having been altered by the presence of life so it now contains 21% oxygen - which humans need to be able to breathe! It has one natural satellite, the Moon, which is the only large satellite of a terrestrial planet in the Solar System.

The Earth's surface is very young. In the relatively short (by astronomical standards) period of 500,000,000 years or so erosion and tectonic processes destroy and recreate most of the Earth's surface and thereby eliminate almost all traces of earlier geologic surface history (such as impact craters). Thus the very early history of the Earth has mostly been erased. The Earth is 4.5 to 4.6 billion years old, but the oldest known rocks are about 4 billion years old and rocks older than 3 billion years are rare. The oldest fossils of living organisms are less than 3.9

billion years old. There is no record of the critical period when life was first getting started.

71 Percent of the Earth's surface is covered with water. Earth is the only planet on which water can exist in liquid form on the surface (though there may be liquid ethane or methane on Titan's surface and liquid water beneath the surface of Europa). Liquid water is, of course, essential for life as we know it. The heat capacity of the oceans is also very important in keeping the Earth's temperature relatively stable. Liquid water is also responsible for most of the erosion and weathering of the Earth's continents, a process unique in the solar system today (though it may have occurred on Mars in the past).

**Earth from Saturn**

*"Look again at that dot. That's here. That's home. That's us. On it everyone you love, everyone you know, everyone you ever heard of, every human being who ever was, lived out their lives. The aggregate of our joy and suffering, thousands of confident religions, ideologies, and economic doctrines, every hunter and forager, every hero and coward, every creator and destroyer of civilization, every king and peasant, every young couple in love, every mother and father, hopeful child, inventor and explorer, every teacher of morals, every corrupt politician, every "superstar," every "supreme leader," every saint and sinner in the history of our species lived there-on a mote of dust suspended in a sunbeam.          ------------Carl Sagan, Pale Blue Dot*

The Earth's atmosphere is 77% nitrogen, 21% oxygen, with traces of argon, carbon dioxide and water. There was probably a very much larger amount of carbon dioxide in the Earth's atmosphere when the Earth was first formed, but it has since been almost all incorporated into carbonate rocks and to a lesser extent dissolved into the oceans and consumed by living plants. Plate tectonics and biological processes now maintain a continual flow of carbon dioxide from the atmosphere to these various "sinks" and back again. The tiny amount of carbon dioxide resident in the atmosphere at any time is extremely important to the maintenance of the Earth's surface temperature via the greenhouse effect. The greenhouse effect raises the average surface temperature about 35 degrees C above what it would otherwise be (from a frigid -21 C to a comfortable +14 C); without it the oceans would freeze and life as we know it would be impossible. (Water vapor is also an important greenhouse gas.)

The interaction of the Earth and the Moon slows the Earth's rotation by about 2 milliseconds per century. Current research indicates that about 900 million years ago there were 481 18-hour days in a year. Earth has a modest magnetic field produced by electric currents in the outer core. The interaction of the solar wind, the Earth's magnetic field and the Earth's upper atmosphere causes the auroras (see the Interplanetary Medium). Irregularities in these factors cause the magnetic poles to move and even reverse relative to the surface; the geomagnetic North Pole is currently located in northern Canada.

(The "geomagnetic north pole" is the position on the Earth's surface directly above the South Pole of the

Earth's field.) The Earth's magnetic field and its interaction with the solar wind also produce the Van Allen radiation belts, a pair of doughnut shaped rings of ionized gas (or plasma) trapped in orbit around the Earth. The outer belt stretches from 19,000 km in altitude to 41,000 km; the inner belt lies between 13,000 km and 7,600 km in altitude.

**Mars - the red planet**

Mars is smaller than both Earth and Venus. The first spacecraft to visit Mars was Mariner 4 in 1965. Several others followed, most recently in 2008, when Phoenix landed in the northern plains to search for water. Three Mars orbiters (Mars Reconnaissance Orbiter, Mars Odyssey, and Mars Express) are also currently at work studying Mars. NASA has landed several unmanned robotic probes on Mars, most recently two remote controlled car-like robots called Mars Rovers. These probes allow NASA scientists to explore the planet, take pictures, analyses soil and conduct experiments. The picture on the left is of one of the Mars Rovers on the surface of Mars. Mars is named after the Greek God of War. It is sometimes also called the red planet, because most of its surface is covered in reddish rocks, dust and soil. Mars is smaller than Earth and Venus. It has a thin atmosphere that is composed mainly of carbon dioxide.The color of the planet is red due to rust in iron-rich soil. Mars has two small satellites named Dimos and Phobos. Gravity on the surface of Mars is 37% of Earth, which means that you can jump up to three times higher than Mars .The highest mountain in the Solar System on Mars is the Olympus and the huge volcano which is 21 kilometers high and 600 kilometers diameter. According to many scientists, it may still be active. Mars is the largest dust storm and lasts for a month.

Selfie by Curiosity stream

The Mars Science Laboratory and its rover centerpiece, Curiosity, is the most ambitious Mars mission yet flown by NASA. The rover's primary mission is to find out if Mars is, or was, suitable for life. Another objective is to learn more about the red planet's environment. Curiosity's size allows it to carry a host of scientific experiments to zap, analyze and take pictures of any rock within reach of its 7-foot (2 meters) arm. Curiosity is about the size of a small SUV. The rover has a few tools to search for habitability. Primary mission: Can, or could, Mars support life?

**Asteroid**

Asteroids are rocky worlds revolving around the sun that are too small to be called planets. They are also known as planetoids or minor planets. There are millions of asteroids, ranging in size from hundreds of miles to several feet across. In total, the mass of all the asteroids is less than that of Earth's moon. Despite their size, asteroids can be dangerous. Many have hit Earth in the past, and more will crash into our planet in the future. That's one reason scientists study asteroids and are eager to learn more about their numbers, orbits and physical characteristics. If an asteroid is headed our way, we want to know that.

## Meteoroid

Meteoroids are all the smaller objects in orbit around the Sun. Most of them originate from comets that lose gas and dust when they approach the Sun. Other meteoroids are basically small asteroids. This dust particles are running in space. When they enter the Earth's atmosphere, then their speed is more than 70,000 kilometers per hour. Due to the high speed and the friction of the atmosphere, they burn very hot, and its speed creates a flame in the atmosphere of the path. They are called Meteoroid or shooting star. These are actually very small Meteors are burnt to ashes after two or two seconds before they are visible. But the relatively large meteors burn for a long time and look very bright. Much brighter meteors are called Fireballs. Some fireballs are so bright that their presence is visible in the light of day. Some large fireballs enter into the Earth's atmosphere, and due to strong pressure on their inner core, it explode on the sky, But in most cases they are burnt to ashes in the sky.

Sometimes some meteors are so big that they hit the surface before they were completely burned .Since the moon has no protective atmosphere , So thousands of meteorite (and asteroids and comets) coming from space hits the moon surface for millions of years and create lots of crater. There is also a sign of meteorite hit in our earth. In the north of Russia, many people were injured in the recent meteorological attacks. They did not injured by meteorite hit directly .Many people was curious to see the sudden bright light and approach to their window. And the shock wave of the meteorite broke the window glass, this cause the injury. Don't worry there is no possibility of any major meteorite attach soon. NASA scientist are working to identify any risk and planning for defensive mechanism.

Crater by meteor at Arizona

50,000 years ago a fiery giant meteor mass weighing several hundred thousand tons hurtled through space and impacted the earth near Arizona. A huge depression, 0.74 miles in diameter and 550 feet deep was created. In 1953, Claire Cameron examined Meteor and revealed the age around 4.55 billion years old which is same as solar system age.

**Jovian planets**

The four large outer planets Jupiter, Saturn, Uranus, and Neptune are known as the "Jovian planets" (meaning "Jupiter-like") because they are all huge compared to the terrestrial planets, and because they are gaseous in nature rather than having rocky surfaces .Two of the outer planets beyond the orbit of Mars ( Jupiter and Saturn) are known as gas giants. The more distant Uranus and Neptune are called ice giants. This is because, while the first two are dominated by gas, while the last two have more ice. All four contain mostly hydrogen and helium.

**Jupiter**

Jupiter is the biggest planet in the solar system. It is 2.5 times the mass of all the other planets of the solar system put together! It is a gas giant, rather than a terrestrial planet, and is made largely of hydrogen and helium. The large spot on Jupiter is actually a storm that has been raging for several hundred years! Jupiter was first visited by Pioneer 10 in 1973 and later by Pioneer 11, Voyager 1, Voyager 2 and Ulysses. The unmanned spacecraft Galileo orbited Jupiter for eight years. In 2003 Galileo was crashed deliberately into Jupiter, to stop it from impacting on Europa, one of Jupiter's moons that scientists believe may harbor some basic form of life. Jupiter is called a gas giant planet. Its atmosphere is made up of mostly hydrogen gas and helium gas, like the sun. The planet is covered in thick red, brown, yellow and white clouds. The clouds make the planet look like it has stripes.

One of Jupiter's most famous features is the Great Red Spot. It is a giant spinning storm, resembling a hurricane. At its widest point, the storm is about 3 1/2 times the diameter of Earth. Jupiter is very windy. Winds range from 192 mph to more than 400 mph. Jupiter has three thin rings that are difficult to see. NASA's Voyager 1 spacecraft discovered the rings in 1979. Jupiter's rings are made up mostly of tiny dust particles.

**Jupiter and its red spot detail view**

Jupiter's mass is $1.90 \times 10^{27}$ kg which is about 318 times more than the masses of our planet. If its mass was a bit higher, then its gravitational pressure would have made the internal hydrogen atom nuclear fusion to helium. And same time would the photon or light was produced within it. Jupiter would turn into a star. Scientist Cart Sagan called Jupiter a failed star or "a star that failed". He mention in his popular cosmos series that if the mass of Jupiter would be more, we would be a resident of the two solar system, then night would be rare for us.

## What is escape velocity?

If you throw an object straight up, it will rise until the negative acceleration of gravity stops it, then returns it to Earth. Gravity's force diminishes as distance from the center of the Earth increases, however. So if you can throw the object with enough initial upward velocity so that gravity's decreasing force can never quite slow it to a complete stop, its decreasing velocity can always be just high enough to overcome gravity's pull. The initial velocity needed to achieve that condition is called escape velocity. Escape velocity is defined as the smallest speed that we need to give an object in order to allow it to completely escape from the gravitational pull of the planet. From the surface of the Earth, escape velocity (ignoring air friction) is about 7 miles per second, or 25,000 miles per hour. Given that initial speed, an object needs no additional force applied to completely escape Earth's gravity. The escape velocity of Jupiter is 59.5 kms/second, for the sun it is 617.5 kms/s and for the black hole it is more than the speed of light.

**Why we are grateful to Jupiter?**

The formation of solar system was very violent process. Cosmic objects (asteroids, comets) were flying around and they were hitting each other. Only after being calm, the life evolve in the earth. Even after that, many times the cosmic object hit the earth. After all, life has been flourish again from those disaster. This injury could have been more violent, more fierce asteroids could hit the earth if the Jupiter wouldn't be there. Jupiter is the biggest planet of solar system. And whenever an asteroid or comet fly into the solar system. Then gravity of each planets attracted them toward it. Most of them have definitely run toward the Saturn and Jupiter as they have larger gravity due to their giant size. They also got many hit by the asteroid and comments. A large asteroid belt has been formed between Mars and Jupiter, due to the high gravity force of Jupiter.

**Saturn - the ringed planet**

Saturn is distinguished by its extensive ring system, but otherwise has several similarities to Jupiter. They are both gas giants. Saturn has at least sixty known satellites; two of which, Titan and Enceladus, show signs of geological activity, though they are largely made of ice.

Saturn was first visited by NASA's Pioneer 11 in 1979 and later by Voyager 1 and Voyager 2. Cassini (a joint NASA / ESA project) arrived on July 1, 2004 and is still in orbit now. Saturn's rings are extraordinarily thin: though they're 250,000 km or more in diameter they're less than one kilometer thick. The ring particles seem to be composed primarily of water and ice, but they may also include rocky particles with icy coatings.

Though they look continuous from the Earth, the rings are actually composed of innumerable small particles each in an independent orbit. They range in size from a centimeter or so to several meters. A few kilometer-sized objects are also likely. Saturn's rings are extraordinarily thin: though they're 250,000 km or more in diameter they're less than one kilometer thick. Despite their impressive appearance, there's really very little material in the rings -- if the rings were compressed into a single body it would be no more than 100 km across.

**Saturn on the left and Uranus on right**

**Uranus**

Uranus is the lightest of the outer planets, a type of gas giant that some scientists call an ice giant. As you can imagine from this nickname its atmosphere is very cold - the coldest in the solar system. The wind on Uranus can blow at over 500 miles per hour. It was discovered by William Herschel, a famous astronomer, while systematically searching the sky with his telescope on March 13, 1781. Uranus has been visited by only one spacecraft, Voyager 2 on Jan 24 1986. The picture on the left is an enhanced image of Uranus that was beamed back to Earth by Voyager 2. Most of the planets spin on an axis nearly perpendicular to the plane of the ecliptic but Uranus' axis is almost parallel to the ecliptic.

The first planet discovered by telescope. It takes 84 years to orbit once in the sun .His own axis leaned more than 90°. Therefore, ne part of in the dark for 42 years or its night is 42 years long and the day is 42 years long Uranus is mostly hydrogen but it has methane. Methane absorbs red light and the blue light radiates the planet, showing the blue

**Neptune**

Neptune is the outermost planet of the solar system. It is slightly smaller than Uranus. Neptune has also been visited by only one spacecraft, Voyager 2, on Aug 25 1989. Neptune has a mark on it that looks very similar to Jupiter's great spot. Just like Jupiter, this is caused by violent storms. The weather is very extreme on Neptune - the wind on Neptune is the strongest on any planet, and blows at 1,300 miles per hour - as fast as a jet fighter plane. The size of Neptune is less than the Uranus, but the mass is more than that. Uranus mass is 14 times the mass of Earth and the mass of Neptune is 17 times. That's why Neptune's density is relatively high. It radiates more heat, but this radiation is less than Jupiter or Saturn. The number of Neptune's known satellite is 13. Among them, the largest Triton is geologically active. This satellite has hot springs and liquid nitrogen. There are several small asteroids in Neptune's orbits called Neptune Trojans. These Trojans orbit the sun with the mother planet.

Neptune is not found in any ancient civilization because it is not visible to the naked eye. The planet is initially discovered with mathematical calculation later on discovered through telescope. After Jupiter, the solar system has the second largest gravitational force. Neptune's orbit is about 30 astronomical units (AU) from the Sun. That is, it is about 30 times more than the sun's distance from Earth. It will take four hour and six-minute or 246 minute to come a message from Neptune to the earth. That means if you call from Neptune, then it will reach Earth after 246 minutes and you will have to wait another 246 minutes to get an answer.

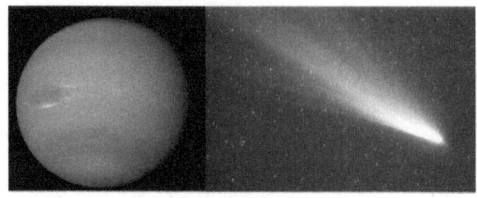

**Neptune on the left and Comets on right**

Comets

Comets are cosmic snowballs of frozen gases, rock and dust roughly the size of a small town. When a comet's orbit brings it close to the sun, it heats up and spews dust and gases into a giant glowing head larger than most planets. The dust and gases form a tail that stretches away from the sun for millions of kilometers. Comets may not be able to support life themselves, but they may have brought water and organic compounds -- the building blocks of life -- through collisions with Earth and other bodies in our solar system.

## Pluto and the other dwarf planets

Pluto used to be classed as a planet of the solar system, but is now considered to be a dwarf planet, and a part of the Kuiper belt. The Kuiper belt is a vast collection of dwarf planets, asteroids, rocks, ice and dust that circle the sun that extends for millions of miles beyond Neptune, on the outskirts of the solar system. As of mid-2008, five smaller objects are classified as dwarf planets, all but the first of which orbit beyond Neptune. These are: Ceres, Pluto, Haumea, make make, Eris

## Reference Data Summary

| Name of Planet | Average Distance from Sun | Diameter | Time to Spin on Axis (a day) | Time to Orbit Sun (a year) | Gravity (Earth = 1) | Average Temperature | Contents of Atmosphere | Number of Known Moons |
|---|---|---|---|---|---|---|---|---|
| Mercury | 57,900,000km (36,000,000 miles) | 4,878km (3,031 miles) | 59 days | 88 days | 0.38 | -183°C to 427°C (-297°F to 800 °F) | Sodium, helium | 0 |
| Venus | 108,160,000km (67,000,000 miles) | 12,104km (7,521 miles) | 243 days | 224 days | 0.9 | 480 °C (896 °F) | Carbon Dioxide (96%), Nitrogen (3.5%) | 0 |
| Earth | 149,600,000km (92,960,000 miles) | 12,756km (7,926 miles) | 23hours, 56 mins | 365.25 days | 1 | 14 °C (57 °F) | Nitrogen (77%), Oxygen (21%) | 1 |

| | | | | | | | | |
|---|---|---|---|---|---|---|---|---|
| Mars | 227,936,640km (141,700,000 miles) | 6,794km (4,222 miles) | 24 hours, 37 mins | 687 days | 0.38 | -63 °C (-81 °F) | Carbon Dioxide( 95.3%), Argon | 2 |
| Jupiter | 778,369,000km (483,500,000 miles) | 142,984km (88,846 miles) | 9 hours, 55 mins | 11.86 years | 2.64 | -130 °C (-202 °F) | Hydrogen, Helium | 66 |
| Saturn | 1,427,034,000 km(888,750,000 miles) | 120,536km (74,900 miles) | 10 hours, 39 mins | 29 years | 1.16 | -130 °C (-202 °F) | Hydrogen, Helium | 62 |
| Uranus | 2,870,658,186km (1,783,744,300 miles) | 51,118km (31,763 miles) | 17 hours, 14 mins | 84 years | 1.11 | -200 °C (-328 °F) | Hydrogen, Helium, Methane | 27 |
| Neptune | 4,496,976,000km (2,797,770,000 miles) | 49,532km (30,779 miles) | 16 hours, 7 mins | 164.8 years | 1.21 | -200 °C (-328 °F) | Hydrogen, Helium, Methane | 13 |

## How sun effects our life?

Earths has two source of energy source. One is the energy trapped inside it during formation and other is the sun. From the sun Energy comes to Earth in the form of radiation energy or sun ray. We see some light rays. And most of the light rays cannot be seen in our eyes. Like the other stars in the universe, the sun also a burning gas sphere. It radiates huge amount of energy every day. Very small amount is coming to earth. Most rays are spread out into space. Most of these rays are reflected back from the outer atmosphere of the Earth before reaching the earth. And the rest comes into the earth and it absorbs by the earth and turn into heat. This heat keeps the earth worm, keeps us alive. We can't live without the sun energy. Without the sun would freeze to death. All life can survive in earth because it does have perfect temperature to keep life working. The sun energy turns into heat and keeping us worm, the plants are making food for us. More energy gets released from the sun than the planet earth receives solar energy in a year. This energy directly comes from the sun. The sun is like other star, a burning sphere of hydrogen and helium. Only one part of two billionth parts of the energy radiated Earth gets. Yet the amount of energy is huge, the energy the whole earth consume in a year can be produce by sun in a few minutes.

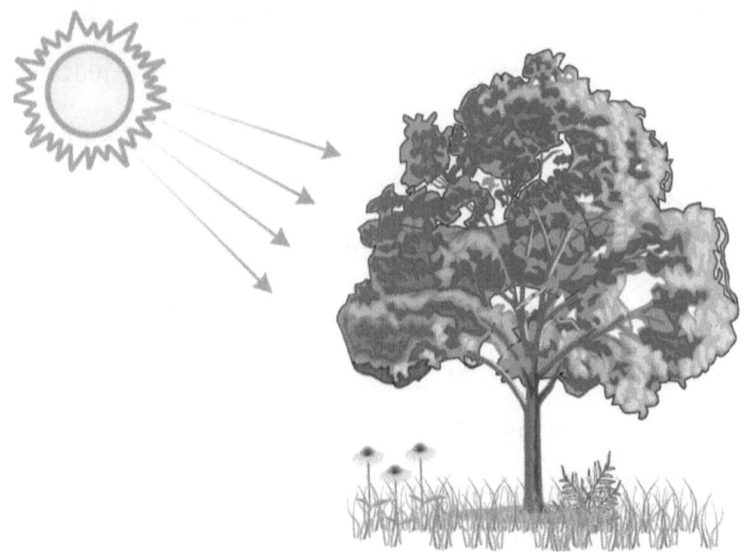

Image Source: Awesome stuff by C. Broman

**What is Photosynthesis?**

The plants use the sunlight to prepare their food by transferring water and carbon di oxide it into sugar. This provide energy to the plants to grow. Plants store this energy into their stems, fruits, and roots, and in the leaves. The conversion of sunlight energy into usable chemical energy, is associated with the actions of the green pigment is called photosynthesis.

Image Source: Awesome stuff by C. Broman

The fish and meat that we eat also comes from the plants. All animals, including fish, cows and goats eats plants and they store energy in their body. We used those as food and our body system convert those energy into heat energy to keep us worm and muscle energy to move and electromagnetic energy to think. So we are all connected to sun and earth, all life are interconnected. All the elements used to build our body are been collected from blood and breathing. Those elements was born into distance old star that dies.

*"The nitrogen in our DNA, the calcium in our teeth, the iron in our blood, the carbon in our apple pies were made in the interiors of collapsing stars. We are made of star stuff."*
― Carl Sagan, Cosmos

## What is the cause of diversity?

How sun create the weather and diverse life in earth?

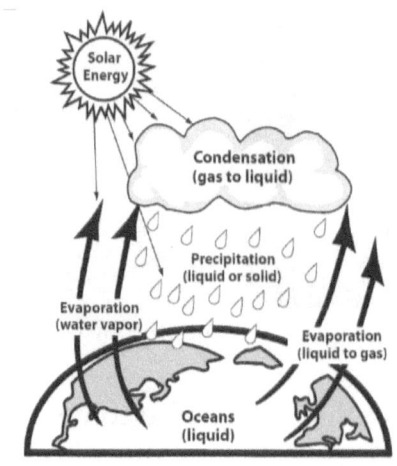

Weather processes such as wind, clouds, and precipitation are all the result of the atmosphere responding to uneven heating of the Earth by the Sun. The uneven heating causes temperature differences, which in turn cause air currents (wind) to develop, which then move heat from where there is more heat (higher temperatures) to where there is less heat (lower temperatures). The atmosphere thus becomes a giant "heat engine", continuously driven by the sun. High and low pressure areas, wind, clouds, and precipitation systems are all caused, either directly or indirectly, by this uneven heating and the resulting heat redistribution processes.

Air naturally moves from high to low pressure and cause wind. We can use the energy on wind such as windmill, sailboat etc. The energy in wind comes from the sun. When the sun shines, some of its light reaches the Earth's surface. The Earth near the Equator receives more of the sun's energy than the North and South Poles. Some parts of the Earth absorb more solar energy than others. Some parts reflect more of the sun's rays back into the air. Light-colored surfaces and water reflect more sunlight than dark surfaces. Snow and ice reflect sunlight, too. When the Earth's surface absorbs the sun's energy, it turns the light into heat. This heat on the Earth's surface warms the air above it. The air over the Equator gets warmer than the surface air near the poles. The air over the desert gets warmer than the air in the mountains. The air over the land usually gets warmer than the air over the water. As air warms, it expands. The warm air over the land becomes less dense than the cooler air and rises into the atmosphere. Cooler, denser air nearby flows in to take its place. This moving air is what we call wind. It is caused by the uneven heating of the Earth's surface.

**Food chain or cycle**

Sun is the source of all energy including the food chain of earth and life. Among the all life on earth only the plants can store suns energy. The plants use their leaves to produce energy, the process called photosynthesis. All other life directly or indirectly depends on plants. Animals get energy from the food they eat, and all living things get energy from food. Food chains begin with plant-life, and end with animal-life. Some animals eat plants, some animals eat other animals. The food we consume is the source of energy and we use this to control out body movement, to see, to hear, to taste, to smell, and to think. All these energy is coming from the sun.

**Ultraviolet Ray**

Electromagnetic radiation comes from the sun and transmitted as waves or particles at different wavelengths and frequencies. Sunlight waves have different web lengths and frequency known as electromagnetic spectrum. Electromagnetic spectrum higher frequency in the visible spectrum is violet.

Protection layer

Just after violet the ultraviolet frequency energy also comes to earth from the sun. This small wave length radiation is ultraviolet radiation. Exposure to **ultraviolet (UV) radiation** is a major risk factor for most skin cancers. Using sun protection sun screen, sunglasses and umbrella is protective measure. The ozone layer or ozone shield is a region of Earth's stratosphere that absorbs most of the Sun's ultraviolet (UV) radiation. It contains high concentrations of ozone gas, although still small in relation to other gases in the stratosphere. Using bio fuel and increase industrial air pollution increase carbon oxide which is reducing the ozone layer. This is why more ultraviolet energy in entering into earth and making the earth hot. As a result the polar ice is melting and sea level is rising, if this continues some coastal countries and cities will be under water and lots of people will lose their home. A lots of food growing land will lose food production capacity due to salty water. We should take care our earth because we have only one home that is our earth.

**The power of Sun**

We use gas or bio fuel for car, airplane and produce electricity. Why don't we use all our energy from the sun? The reason is that we don't know how to collect the sun energy efficiently. The solar cell we have now can collect very little amount of energy from the sun and convert into electric energy. Photovoltaic (PV) cells

Converts solar power into electricity. The photo mean light and the voltage is the unit of measurement of electricity pressure. PV cells are made to be silicon. There is a chemical substance on the silicon biscuits. Chemical substance (BORON) produces electricity when the light energy is hits on the photovoltaic (PV) cells. Today we can convert very small amount of solar power but In the future it may be a great source of energy. Scientists are always looking for new ways to use solar power.

## What is goldilocks zone?

The goldilocks zone is also know as habitable zone. Mercury and Venus were too hot to support life. Mercury even does not have atmosphere and are vulnerable for extreme solar wind. Venus is the hottest planet due to dense atmosphere of acid and carbon di oxide. There is no sign of any life in mars. The outer planets were too cold. Only Earth is just right orbit and location and just right for life. Our planet has liquid water, a breathable atmosphere, a suitable amount of sunshine. Perfect, weather variation due to the right tilt provide diverse geography hence a diverse life. If Earth were a little closer to the sun it might be like hot choking Venus; a little farther, like cold arid Mars. Somehow, though, we ended up in just the right place with just the right ingredients for life to flourish. Researchers of the 1970s scratched their heads and said we were in "the Goldilocks Zone."

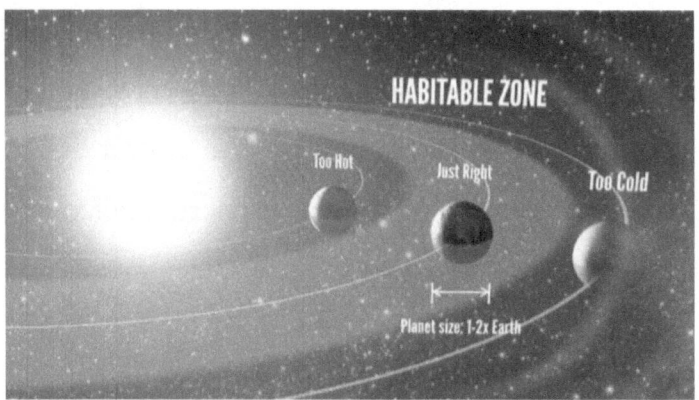

The Goldilocks Zone seemed a remarkably small region of space. It didn't even include the whole Earth. All life known in those days was confined to certain limits: no colder than Antarctica (penguins), no hotter than scalding water (desert lizards), no higher than the clouds (eagles), no lower than a few mines (deep mine microbes). In the past 30 years, however, our knowledge of life in extreme environments has exploded. Scientists have found microbes in nuclear reactors, microbes that love acid, microbes that swim in boiling-hot water. Whole ecosystems have been discovered around deep sea vents where sunlight never reaches and the emerging vent-water is hot enough to melt lead. Even scientist send a small microbe known as water bear (tardigrade) to international space station. It was exposed to high radiation of sun and found it still survive. So the Goldilocks Zone is bigger than we thought. To find out how big, researchers are going deeper, climbing higher, and looking in the nooks and crannies of our own planet. Searching for life in the Universe is one of NASA's most important research activities. Finding extreme life here on Earth tells us what kind of conditions might suit life "out there."

**Is there any life out there?**
Is there any goldilocks zone or habitable zone elsewhere?

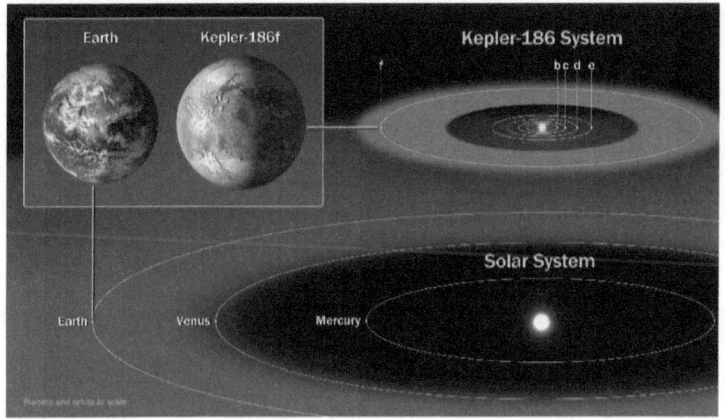

The artist's concept depicts Kepler-186f, the first validated Earth-size planet to orbit a distant star in the habitable zone *Image: NASA Ames/SETI Institute/JPL-Caltech*

Astronomers have added few hundred candidates of goldilocks zone in other solar system. And the list of planets beyond our solar system is growing. Recent 10 of which may be about the same size and temperature as Earth, boosting their chances of hosting life. NASA's Kepler Space Telescope, astronomers have discovered the first Earth-size planet orbiting a star in the "habitable zone" -- the range of distance from a star where liquid water might pool on the surface of an orbiting planet. The discovery of Kepler-186f confirms that planets the size of Earth exist in the habitable zone of stars other than our sun. While planets have previously been found in the habitable zone, they are all at least 40 percent larger in size than Earth and understanding their makeup is challenging. Kepler-186f is more reminiscent of Earth. NASA launched the Kepler telescope in 2009 to learn if Earth-like planets are common or rare. With the final analysis of Kepler data in hand, scientists said they will now work on answering that question, a key step in assessing the chance that life exists beyond Earth.

Recently NASA's Kepler discovers 10 new Earth-sized, potentially habitable planets

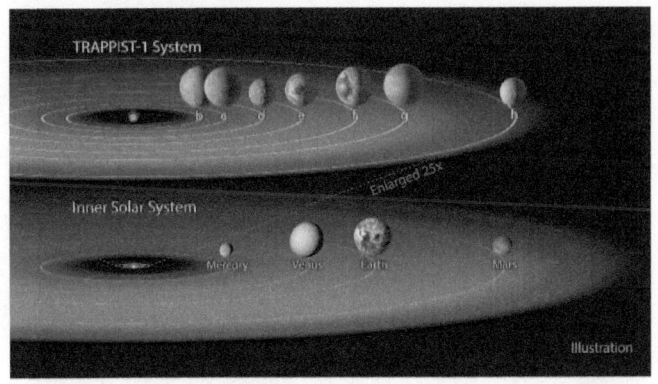

The TRAPPIST-1 system contains a total of seven planets, all around the size of Earth. Three of them -- TRAPPIST-1e, f and g -- dwell in their star's so-called "habitable zone. "The habitable zone, or Goldilocks zone, is a band around every star (shown here in green) where astronomers have calculated that temperatures are just right -- not too hot, not too cold -- for liquid water to pool on the surface of an Earth-like world.

## How do we know all this?

Scientist use *scientific method* to prove their claim based on evidence. *scientific method* include an observation, asking a question, Form a hypothesis in a testable form, make a prediction based on the hypothesis and Test the prediction. And scientist use the results to make new hypotheses or predictions and release those information for peer review. Astronomer use telescope for observation, some small scale test on particle in done at laboratory to establish the theory. The big bang theory and particle standard model are the biggest one in our modern science. There are few information is given below about those tools.

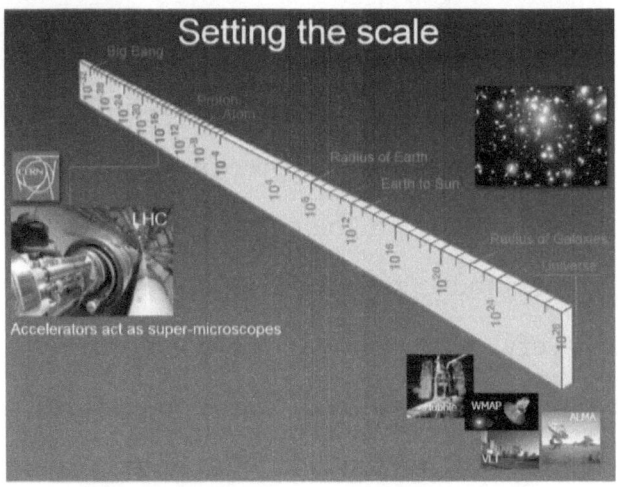

Scientist build amazing instrument, satellite and telescope to know the story from big Bang to present.

**What is the International Space Station (ISS)?**

Do people live in the International Space Station? International Space Station is a large spacecraft. It orbits around the earth and it is 220 mile above the earth surface. It is big like six bedroom house where the astronauts live. There is a science lab in the space station Where scientist do various research such as how different things work in zero gravity, how water behaves in zero gravity, how bacteria lives in high radiation. Many countries have built it together and they all work together. Since it is not possible to take such a big structure at once from the earth, so the space station is made up of several module and assembled in the space. The astronauts put the pieces together in the space of day by day. During the time of build once some tools flew into space. Scientist uses the space station to learn how to live and how to work in space such that human can travel in deep space. This is the way scientist is learning to make bigger spacecraft and send people to outer space that will help in exploring the space.

**How do astronauts live in space?**

When the astronauts go to space, they have to wear a special suit. It protects them from the sun's harmful rays and the dust of the space. This special suite has twelve layers and is full of breathable air inside. Since the space is completely empty, this suite gives them earth like air pressure. Without this suite, the oxygen will be released from the body, due to the lack of air pressure, the nose will bleed and the body will cool down. There is a Velcro patch inside an astronaut helmet that helps to scrub the body if they feel itchy. The rovers were taken to the moon was battery powered, as gas powered motor engines will not rub because there is no oxygen in the moon.

**How old is the international space station?**
The first piece of the International Space Station was launched in 1998. A Russian rocket launched that piece. After that, more pieces were added. Two years later, the station was ready for people. In 2nd November 2000, people started living in the first space station. NASA and its partners around the world finished the space station in 2011.

**How Big Is the Space Station?**
The space station is as big inside as a house with five bedrooms. It has two bathrooms, a gymnasium and a big bay window. Six people are able to live there. It weighs almost a million pounds. It is big enough to cover a football field including the end zones. It has science labs from the United States, Russia, Japan and Europe. The astronauts see sunrise and sunset around 15 times daily.

**Facts of International Space Station**

It's bigger than a football field : The total length of the ISS from end to end is about 109 meters (357 feet), longer than both a "proper" football field and an American Football field, while its livable space is roughly equal to a five-bedroom house. 340 people have been to the station: The first mission to the ISS was on 2 November 2000 and since then it has been continuously occupied. 70 manned missions on Space Shuttles and Soyuz spacecraft have flown to the ISS, while over 60 unmanned vehicles have docked with the station. It's the most expensive object ever built: At an estimated cost of $100bn dollars, the ISS is the most expensive single object ever built by mankind. Roughly half of the total price was contributed by the USA, the rest by other nations including Europe, Japan and Russia. It weighs more than 320 cars: The ISS is primarily composed of 15 pressurized modules (seven US, five Russian, two Japanese and one European) and four large solar panels. It weighs 420,000 kilograms (925,000 pounds), which is more than 320 automobiles. Because the human body tends to lose muscle and bone mass in zero gravity environments, all astronauts aboard the ISS must work out at least two hours a day to combat these effects. Oxygen in the ISS comes from a process called "electrolysis," which involves using an electrical current generated from the station's solar panels to split water molecules into hydrogen and oxygen gas. Astronauts urinate in a special tube, which collects water through the water treatment plant to drinkable water. Astronauts ate their own grown lettuce in space station on August 10, 2015. It speed is 4,791 miles per second, which is enough for a moon to go and return in a day.

## What is the Hubble Telescope?

From the dawn of humankind to a mere 400 years ago, all that we knew about our universe came through observations with the naked eye. Then Galileo turned his telescope toward the heavens in 1610. The world was in for an awakening. Telescopes grew in size and complexity and power. Not only that we place telescope outside the earth for better view.

**Hubble Facts**

NASA's Hubble Space Telescope was launched April 24, 1990, on the space shuttle Discovery from Kennedy Space Center in Florida. Hubble has made more than 1.3 million observations since its mission began in 1990. Astronomers using Hubble data have published more than 15,000 scientific papers, making it one of the most productive scientific instruments ever built. Hubble does not travel to stars, planets or galaxies. It takes pictures of them as it whirls around Earth at about 17,000 mph. Hubble has traveled more than 4 billion miles along a circular low Earth orbit currently about 340 miles in altitude. Hubble has no thrusters. To change pointing angles, it uses Newton's third law by spinning its wheels in the opposite direction. It turns at about the speed of a minute hand on a clock, taking 15 minutes to turn 90 degrees. Hubble has the pointing accuracy of .007 arc seconds, which is like being able to shine a laser beam focused on Franklin D. Roosevelt's head on a dime roughly 200 miles away. Outside the haze of our atmosphere, Hubble can see astronomical objects with an angular size of 0.05 arc seconds, which is like seeing a pair of fireflies in Tokyo from your home in Maryland. Hubble has peered back into the very distant past, to locations more than 13.4 billion light years from Earth. The Hubble archive contains more than 140 terabytes, and Hubble science data processing generates about 10 terabytes of new archive data per year. Hubble weighed about 24,000 pounds at launch and currently weighs about 27,000 pounds following the final servicing mission in 2009 – on the order of two full-grown African elephants. Hubble's primary mirror is 2.4 meters (7 feet, 10.5 inches) across. Hubble is 13.3 meters (43.5 feet) long -- the length of a large school bus.

Ref: https://www.nasa.gov/mission_pages/hubble/story/index.html

## What is the Large Hadron Collider (LHC)?

The Large Hadron Collider (LHC) is the world's largest and most powerful particle accelerator. It first started up on 10 September 2008, and remains the latest addition to CERN's accelerator complex. The LHC consists of a 27-kilometre ring of superconducting magnets with a number of accelerating structures to boost the energy of the particles along the way. It was built by the European Organization for Nuclear Research. It is a giant circular tunnel built underground. A hadron is a particle which consists of a number of quarks held together by the subatomic strong force. This powerful accelerator is used to test and validate the standard model of particle physics. The Standard Model of particle physics is a theory developed in the early 1970s that describes the fundamental particles and their interactions. It has precisely predicted a wide variety of phenomena and so far successfully explained almost all experimental results in particle physics. But the Standard Model is incomplete.

Image of LHC

It leaves many questions open, which the LHC will help to answer. How does the quark-gluon plasma give rise to the particles that constitute the matter of our Universe? What is the origin of mass? Why is there far more matter than antimatter in the universe? What are dark matter and dark energy? The collider will examine its immense new domain for evidence of hidden spacetime dimensions, new strong interactions, supersymmetry and the totally unexpected. It is a machine to produce a condition of big bang in a miniscule to understand how the matter form from the energy.

https://home.cern/topics/large-hadron-collider

## How to map the universe?

To map the universe scientist attempt different mission of satellite such COBE, PLANK and WMAP. The Planck mission detected the oldest light in our universe with the greatest precision. The ancient light, called the cosmic microwave background, was imprinted on the sky when the universe was 370,000 years old. It shows tiny temperature fluctuations that correspond to regions of slightly different densities, representing the seeds of all future structure: the stars and galaxies of today. By analyzing the light patterns in this map, scientists are fine tuning what we know about the universe, including its origins, fate and basic components.

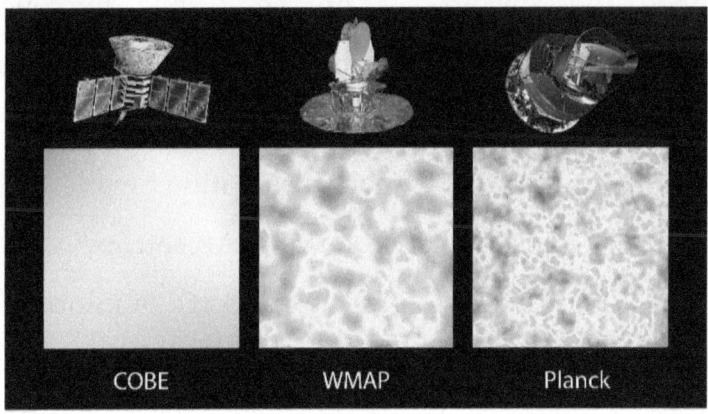

The Cosmic Background Explorer **(COBE),** was a satellite dedicated to cosmology. It operated from 1989 to 1993. Its goals were to investigate the cosmic microwave background radiation (CMB) of the universe and provide measurements that would help shape our understanding of the cosmos.

The Wilkinson Microwave Anisotropy Probe (**WMAP**) is a NASA Explorer mission that launched June 2001 to make fundamental measurements of cosmology to the study of the properties of our universe as a whole.

**Planck** was a space observatory operated by the European Space Agency (ESA) from 2009 to 2013, which mapped the anisotropies of the cosmic microwave background (CMB) at microwave and infra-red frequencies, with high sensitivity. The mission substantially improved upon observations made by the NASA Wilkinson Microwave Anisotropy Probe (WMAP). Planck provided a major source of information relevant to several cosmological and astrophysical issues, such as testing theories of the early Universe and the origin of cosmic structure; as of 2013, it has provided the most accurate measurements of several key cosmological parameters, including the average density of ordinary matter and dark matter in the Universe.

## What is James Webb Telescope (JWST)?

The James Webb Space Telescope will be launched on an Ariane 5 rocket on spring 2019. Partnership JWST is a joint project between NASA, ESA and the Canadian Space Agency (CSA). JWST is a major space observatory often presented as the successor to the very successful NASA/ESA Hubble Space Telescope (HST). It has a large 6.5-metre segmented mirror that will collect almost 6 times more light than HST. It has also been designed to work with infrared light. It will address a broad variety of scientific topics ranging from detecting the first galaxies in the Universe to studying planets around other stars. JWST will carry sophisticated Instruments and camera. It will be able to see further deep and older light than Hubble telescope. It will be looking mostly in the infrared but also some in the red part of the visible light (the pictures will be color coded so we can see them). It will be able to see things that the Hubble Space Telescope cannot. Infrared vision can be used to see heat (like some kinds of night vision goggles).

New James Webb telescope (image:NASA)

After launch and it will travel 1 month on a transfer trajectory, the observatory will operate at approximately 1.5 million kilometers from Earth, in an orbit around the second Lagrange point of the Sun-Earth system.

## Finally a Mysterious universe revealed!!!

From history, it is seen that mankind has always imagined himself in the center of the universe. Even Isaac newton also thoughts we are at the centre of the universe. Human always wondering about the origin and formation of the universe. It was Nicolaus Copernicus who bring the idea that we are not at the center of the universe, we are just a planet of our sun. And Galileo proved his idea by his telescope. But so far all astronomical evidence has been collected, with the help of modern technology how the universe has been observed, with the development of math, physic, geology and other science. This earth or sun centric concept has been abandoned. There is no center of the universe, we are not in special place into the vast universe. Our earth is like a speck of dust in cosmic scale. There huge and interesting cosmic objects are floating like an island in the dark ocean of cosmos. Today we know that we are living in the planet that is average in size and belongs to an average star size family. And our solar system (sun with its family) orbits around Milky Way galaxy. There are few hundred billion solar system exist in our Milky Way galaxy. And there are billions of galaxy in our visible universe. Each of them are running at hundreds of kilometers per second toward unknown destination. The universe is not static, it is expanding. Who know what is swimming under the ocean of Saturn's moon, who knows what other form of life is evolving out there. Who knows how many more intelligent civilization is forming in other planets. Cosmology is a branch of science that works with the origin and formation of all large structure of the cosmos and the universe itself. How the underlying process works, scientist try to have the deep understanding the laws of physics. The whole universe is govern by laws of physics .The laws of physics we observed must be same throughout the universe. As we live in the cosmos, so we can't do the experiment with the whole universe but we can observe and study the cosmos. We do experiment with each small part and collectively we understand the whole. We can't create big bang but scientist use to simulate this in a small particle level at large hadron collider. Our science will only can know about the universe we live in. If there any parallel universe we will never come to know how that universe works. It might be different laws of physics govern that universe. The cosmos is unimaginable huge and does have long history. We might not find all the signs from his history very easily. But scientist like newton, Einstein, Sathandranath Bosh, Carl Sagan, Richard Feynman all mathematician, physicist gave and giving us new way to look about the cosmos. Galileo to Hubble to James telescopes are our eyes in space. In cosmic period of 13.8 billion years, our life even the whole human civilization is very short, just like blink of eyes. And modern science is just a few hundred years. But we are lucky to live in scientific era where we can know the about the cosmos with eye witness evidence. We are coming to know new stuff about the cosmos every day. Thanks all scientists, engineers and astronomer. I feel very successful that you read this book. I hope you have a better understanding about the depth and the vastness of the universe and same time raises many more question. Hope your curiosity will drive you to learn more about cosmos.

www.ingramcontent.com/pod-product-compliance
Lightning Source LLC
Chambersburg PA
CBHW021410210526